Modeling and Simulation in Science, Engineering and Technology

Series editor
Nicola Bellomo
Politecnico di Torino
Torino, Italy

Editorial Advisory Board

For further volumes:
http://www.springer.com/series/4960

Natalia L. Komarova · Dominik Wodarz

Targeted Cancer Treatment in Silico

Small Molecule Inhibitors and Oncolytic Viruses

 Birkhäuser

Natalia L. Komarova
Department of Mathematics
University of California, Irvine
Irvine, CA
USA

Dominik Wodarz
Department of Ecology and
   Evolutionary Biology
University of California, Irvine
Irvine, CA
USA

ISSN 2164-3679   ISSN 2164-3725 (electronic)
ISBN 978-1-4614-8300-7   ISBN 978-1-4614-8301-4 (eBook)
DOI 10.1007/978-1-4614-8301-4
Springer New York Heidelberg Dordrecht London

Library of Congress Control Number: 2013945158

Mathematics Subject Classification (2010): 92D15, 92D25, 60J85, 60J27, 37C75, 34D20, 65P99, 37M99

Printed on acid-free paper

Springer is part of Springer Science+Business Media (www.birkhauser-science.com)

This book is dedicated to the memory
of Leonid Alexandrovich Safronov

Preface

The effective treatment of cancer, resulting in long-term remission, is not yet seen in the majority of patients. Cancer is a complex set of diseases that can occur in a variety of organs, defined by cells that escape regulatory mechanisms and proliferate out of control. During recent years, an ever increasing emphasis has been placed on so-called targeted treatment approaches. In the broadest sense, this includes therapies that specifically attack cancer cells without causing significant harm to healthy tissue. This is in contrast to chemotherapies or radiation therapies which target all dividing cells, leading to severe side effects. Of particular importance, small molecule inhibitors that target specific defects in cancer cells initiate and maintain the disease. Especially, the treatment of chronic myeloid leukemia (CML) has seen enormous benefits from this development. Another treatment approach that targets cancer cells specifically is the use of oncolytic viruses. These are viruses that specifically infect and kill cancer cells but not healthy cells. They can spread throughout the tumor and the aim for the virus is to drive the tumor into remission. Tumor cell specificity comes about by the requirement of certain defects commonly found in cancer cells (such as the absence of functional p53) for productive virus replication. Although more in an experimental state of development so far, virus therapy could have relatively broad applicability to a variety of cancers, since its development does not depend on our understanding of molecular defects that initiate and maintain cancers, which is often relatively poor.

Because of the complexity of the disease, people have come to the realization that new approaches and new ways of thinking would be very beneficial for the further development of therapies. In particular, interdisciplinary approaches are thought to be very valuable. Cancers are very complex systems characterized by a multitude of dynamical interactions, and the evolution of cells from a cooperating tissue cell phenotype toward a selfish malignant phenotype is at the core of the disease. Therefore, insights from population dynamics and evolutionary biology cannot be separated from the biomedical thought process. This in turn is tightly connected to tools from the physical sciences, i.e., to the development of mathematical and computational models of carcinogenesis and cancer therapy. This has also been the focus of our research for the last decade or so, and the aim of this book is to review this body of work and to show how mathematical and computational approaches can provide insight that are biomedically relevant and that could influence the further development of treatment strategies. We do this in the context of targeted treatment approaches, including small molecule inhibitors and oncolytic viruses. Besides reviewing the work and the insights, this book also serves as an introduction to the mathematical methodology that is at the core of this analysis. While the exact form of mathematical models and certain insights might change, depending on future insights, the mathematical methodologies are a fundamental tool that will always be needed to address questions about cancer therapy and resistance against treatments.

This work would not have been possible without the valuable interactions with our colleagues. We would like to especially thank Hung Fan, Steve Frank, Simon Levin, John Lowengrub, as well as members of our groups, especially Allen Katouli and Ignacio Rodriguez-Brenes.

Irvine, CA, USA   Natalia L. Komarova
December 2012   Dominik Wodarz

Contents

Acronyms

ABL	Abelson murine leukemia viral oncogene
APC	Antigen presenting cell
ATP	Adenosine triphosphate
BCR	Breakpoint cluster region
BDF	Best drug first
CML	Chronic myeloid leukemia
CTL	Cytotoxic T-lymphocyte
MOI	Multiplicity of infection
ODE	Ordinary differential equation
PDE	Partial differential equation
SEER	Surveillance epidemiology and end results
WDF	Worst drug first

Chapter 1
Background and Scope of the Book

Abstract This chapter outlines the scope and structure of the book. It gives a brief overview of cancer initiation and development, describes general treatment approaches, and defines targeted therapy, the topic of this book. The structure of the book is explained. The book is divided into two parts covering two different types of treatments that specifically target cancer cells: the treatment with small molecule inhibitors and the use of oncolytic viruses.

Keywords Cancer initiation · Cancer therapy · Targeted therapy · Small molecule inhibitors · Oncolytic viruses

1.1 Cancer Concepts

Multicellular organisms, such as humans, function because the cells of the individual tissues cooperate for the good of the entire organism. This means that cellular proliferation needs to be regulated and that cells cannot proliferate beyond a homeostatic level. This homeostasis (the maintenance of constant cell population levels) is regulated by complex mechanisms including feedback loops [1]. When these regulatory mechanisms break down, cells can start to proliferate out of control, leading to the development of cancer [2]. The development of cancer is a somatic evolutionary process [3–9], and methods of evolutionary theory and ecology can be applied to study many aspects of carcinogenesis and treatment [10]. Healthy cells acquire a number of mutations that allow them to escape regulatory mechanisms. In many cancers, a cell has to accumulate several mutations in order to escape homeostasis, a process called multistep carcinogenesis [11–15]. Different types of mutations can be acquired by cells [16]. Oncogenes are genes that need to be activated in order to drive uncontrolled cellular proliferation. Mutation of one copy of the gene is sufficient. Conversely, some genes ensure homeostatic control by producing factors that repress proliferation. These are called tumor suppressor genes, and both copies have to be

N. L. Komarova and D. Wodarz, *Targeted Cancer Treatment in Silico*,
Modeling and Simulation in Science, Engineering and Technology,
DOI: 10.1007/978-1-4614-8301-4_1, © Springer Science+Business Media New York 2014

deactivated in order to observe an aberrant phenotype. In addition, there are genes that ensure the faithful copying and maintenance of the genome. When they become corrupted, increased mutation rates can be observed, leading to different types of genetic instabilities [16].

1.2 Treatment Concepts

The aim of cancer treatment is to stop uncontrolled growth and induce death of cancer cells, such that remission is observed. This has been subject to intense investigation over several decades, with limited success. The traditional treatment approaches to cancers include chemotherapies and radiation therapies. These approaches are based on damaging the genomes of cells such that they halt division and/or die [16]. Dividing cells are targeted in this approach. While cells in a tumor are certainly proliferating, healthy dividing cells in the body are also affected (such as epithelial tissue), and this can lead to very strong side effects. Cellular responses often rely on intact checkpoints in cells such that apoptosis or cellular senescence is induced, and these checkpoints are often inactivated in advanced tumor cells. Moreover, cells can evolve resistance to treatment in a variety of ways [17, 18], most notably they can become multidrug resistant, e.g., by pumping drugs out of the cells. All these aspects make cancer treatment difficult, with relatively low success rates and large degrees of side effects.

Rather than hitting populations of dividing cells in general, recent efforts have concentrated on finding more specific treatment approaches that target cancer cells but not healthy cells. These are called "targeted treatment approaches". In this respect, small molecule inhibitors have recently started to play an important role in the clinical setting. Small molecule inhibitors specifically target molecular defects in cancer cells which eliminates their ability to grow out of control. In order to systematically develop such drugs, one needs to first gain an understanding of the molecular mechanisms that drive and maintain the tumor, and then to design inhibitors to specifically disable these mechanisms. The development of tyrosine kinase inhibitors has been a good example of this, especially in the context of leukemia [19]. The treatment of chronic myeloid leukemia with the targeted inhibitor imatinib and subsequent inhibitors has seen the greatest success so far. Tumor cells are selectively affected, leading to good treatment responses with relatively few side effects. However, with many cancer types, our understanding of the defects that drive and maintain aberrant proliferation remain poorly defined, thus preventing the development of targeted inhibitors. In this case, an alternative approach for targeting cancer cells exists, although this approach is much less developed compared to tyrosine kinase inhibitors. This approach relies of the existence of viruses that replicate specifically in cancer cells, and fail to replicate in healthy cells. These viruses can spread from one tumor cell to another and kill them. Such oncolytic viruses can in principle spread through the whole tumor and drive it into remission [20–22]. Some animal viruses replicate naturally in human cancer cells but not in normal human cells [23, 24]. The vast majority of oncolytic viruses, however, is engineered to achieve tumor specificity. Typically, cellular checkpoints that tend to be inactivated in tumor cells shutdown viral replication in healthy cells. The advantage of this approach is that in principle, tumor cells can be targeted specifically without the need to understand the details of the molecular defects that drive the particular tumor. The downside of this approach is that the correlates of successful treatment are currently not fully understood, thus limiting practical use at this point in time. Several encouraging treatment results have been reported, but consistent success remains elusive.

1.3 Book Structure

This book analyzes these modern, targeted treatment approaches with the use of mathematical and computational models.

The first part of the book focuses on cancer treatment with targeted small molecule (tyrosine kinase) inhibitors. They are increasingly used on the treatment of several tumors, the most successful example being the treatment of CML with tyrosine kinase inhibitors. A major obstacle to success is the presence of drug-resistant mutants. Mathematical models can serve as a useful tool for investigating the evolutionary dynamics of cancer cells, and thus for illuminating the basic principles according to which resistant mutants emerge. This knowledge can in turn be used to design treatment regimes to overcome drug resistance, ensuring longer term success of therapy.

The second part of the book concentrates on oncolytic virus therapy. Mathematical models can be very helpful to complement experimental and clinical data in elucidating the correlates of treatment success. They enable us to understand the dynamics between a growing tumor cell population and a replicating virus population. These dynamics are complex and highly nonlinear, limiting our ability to obtain insights from verbal and graphical reasoning. Mathematical models capture a defined set of biological assumptions and follow them to their precise logical conclusions, giving rise to insights that cannot be obtained by experimentation alone.

Both parts of the book start with a brief biological introduction and overview of the topic. This is not meant to be comprehensive. Instead, it is meant to provide the biological knowledge that is necessary to follow the development of the mathematical models that are the focus of this book. For a detailed biological introduction to small molecule inhibitors and oncolytic virus therapy, the reader is referred to various books, reviews, and primary papers, many of which are cited throughout our book. The first part of the book, concerned with small molecule inhibitors, is the mathematically more complex one. One of the chapters (Chap. 4) is devoted to methodology that is required for the analysis of evolutionary dynamics. In addition, most of the chapters include a self-contained Methodology section that explains further mathematical techniques required to arrive at the results described in the appropriate chapter. These more technical parts of the book can be skipped by readers who are mostly interested in the biological message. The gist of the theoretical message is always explained in a nontechnical language, and related to biological examples.

In order to follow all the technical details, the readers should have a background in basic applied mathematics, such as basic theory and linear stability analysis of ordinary differential equations, the method of characteristics for first-order partial differential equations, some basic probability theory and Markov processes. Advanced undergraduate and graduate students of applied mathematics will be able to follow all the details. Again, readers with a biological background can skip these more technical parts without losing the ability to understand the biological conclusions, which are described in less technical terms.

This book can be used as a text for an upper-division undergraduate or graduate course on mathematical modeling of cancer treatment. At the end of each chapter, we provide a number of problems and research projects. The problems usually are aimed at an in-depth understanding of the mathematical tools used to develop the theory. There are also two kinds of Research projects: some require a more extensive mathematical development (with references provided), and others in-depth research into a biological topic. Finally, there is a number of Numerical projects, which require computer programming and guide a reader in a step-by-step fashion through numerical analysis of treatment strategies.

1.4 Mathematical Models

A wide variety of mathematical methods and models have been used to study cancer and its treatment. The choice of the mathematical methodology is dictated by the biological questions that are being asked, and this also applies to the material covered in this book. With respect to small molecule inhibitors, a large focus is on the evolutionary dynamics of drug-resistant mutants. Because this work is often concerned with the dynamics at low abundances, stochastic birth-death processes are required here. Much attention will be given to this methodology. When dynamics at larger population sizes are considered, ordinary differential equations are often the tool of choice. This applies especially to the oncolytic virus section of this book, where the dynamics of virus spread in established tumors is examined. Spatial models can play an important role since tumors are characterized by intricate spatial structures, and we will consider agent-based spatial models in this respect.

This is obviously not an exhaustive modeling repertoire and omits model types that have been used in other aspects of tumor modeling, such as biophysical models, partial differential equations (PDEs), and hybrid models [25–49]. Multiscale modeling is a tool that is capable of combining great biological complexity in a single model, where different processes occur on vastly different spatial and temporal scales. In this approach, spatial cellular dynamics are often described by agent-based models or cellular automata, while populations that can diffuse over larger distances (such as virus particles, nutrients, and signaling molecules) are described by PDEs. A nice review of this approach is provided in [50].

In this book, we limit ourselves to models where the detailed dynamics are happening on a single temporal and spatial scale. We base our models on evolutionary theory and population dynamics approaches, and consider the collective action of changing populations of cells under different environments. In mathematical modeling in biology, one always faces a certain trade-off between the complexity of the model and its amenability to understanding. In this book, we choose the "reductionist" path. The models we construct and explore are abstractions from reality which aim to grasp only the most important components of a complex system, which are thought to give rise to an observed behavior/phenomena. The resulting transparency and amenability to analysis allows ready comparison with experimental data. Moreover, it allows complete understanding of the mathematical consequences of all the biological assumptions that we input in the model. Observing a given model outcome can always be traced back to the underlying model components. What aspects of the system design give rise to each of the predicted behavior patterns? What will be the consequence of alternative assumptions? Is a given mechanism enough to explain an observation?

From a biological viewpoint, simpler, minimally parameterized models rely on fewer assumptions and formulations of uncertain biological nature. This facilitates comparison with experimental data in relatively simple, in vitro settings, which is a necessary first step in model validation. The obvious next step is to add layers of complexity in an incremental fashion. Once complete understanding of the simpler models is achieved, much progress can be made with more complex scenarios. Our hope is that this book can serve as a platform for further, more complex and more advanced modeling efforts.

References

1. Lander, A., Gokoffski, K., Wan, F., Nie, Q., Calof, A.: Cell lineages and the logic of proliferative control. PLoS Biol **7**(1), e1000015 (2009)
2. Rodriguez-Brenes, I.A., Komarova, N.L., Wodarz, D.: Evolutionary dynamics of feedback escape and the development of stem-cell-driven cancers. Proc. Natl. Acad. Sci. U S A **108**(47), 18983–18988 (2011)
3. Moolgavkar, S.H., Knudson, A.G.J.: Mutation and cancer: a model for human carcinogenesis. J. Natl. Cancer Inst. **66**(6), 1037–1052 (1981)
4. Gatenby, R., Vincent, T.: An evolutionary model of carcinogenesis. Cancer Res. **63**(19), 6212–6220 (2003)
5. Calabrese, P., Tavare, S., Shibata, D.: Pretumor progression: clonal evolution of human stem cell populations. Am. J. Pathol. **164**(4), 1337–1346 (2004)
6. Wodarz, D.: Somatic evolution of cancer cells. Semin. Cancer Biol. **15**(6), 421–422 (2005)
7. Nowak, M.: Evolutionary Dynamics: Exploring the Equations of Life. Belknap Press, Cambridge (2006)
8. Vincent, T., Gatenby, R., et al.: An evolutionary model for initiation, promotion, and progression in carcinogenesis. Int. J. Oncol. **32**(4), 729 (2008)
9. Greaves, M., Maley, C.: Clonal evolution in cancer. Nature **481**(7381), 306–313 (2012)
10. Pienta, K., McGregor, N., Axelrod, R., Axelrod, D.: Ecological therapy for cancer: defining tumors using an ecosystem paradigm suggests new opportunities for novel cancer treatments. Trans. Oncol. **1**(4), 158 (2008)

11. Nordling, C.: A new theory on the cancer-inducing mechanism. Br. J. Cancer **7**(1), 68 (1953)
12. Armitage, P., Doll, R.: The age distribution of cancer and a multi-stage theory of carcinogenesis. Br. J. Cancer **8**(1), 1 (1954)
13. Moolgavkar, S.H.: The multistage theory of carcinogenesis and the age distribution of cancer in man. J. Natl. Cancer Inst. **61**(1), 49–52 (1978)
14. Luebeck, E.G., Moolgavkar, S.H.: Multistage carcinogenesis and the incidence of colorectal cancer. Proc. Natl. Acad. Sci. U S A **99**(23), 15095–15100 (2002)
15. Moolgavkar, S.H., Luebeck, E.G.: Multistage carcinogenesis and the incidence of human cancer. Genes Chromosomes Cancer **38**(4), 302–6 (2003)
16. Kinzler, K.W., Vogelstein, B.: The Genetic Basis of Cancer. McGraw-HIll, Toronto (1998)
17. Goldie, J.H., Coldman, A.J.: Drug Resistance in Cancer: Mechanisms and Models. Cambridge University Press, Cambridge (1998)
18. Kruh, G.D.: Introduction to resistance to anticancer agents. Oncogene **22**(47), 7262–4 (2003)
19. Zhang, J., Yang, P., Gray, N.: Targeting cancer with small molecule kinase inhibitors. Nat. Rev. Cancer **9**(1), 28–39 (2009)
20. Davis, J.J., Fang, B.: Oncolytic virotherapy for cancer treatment: challenges and solutions. J. Gene Med. **7**(11), 1380–1389 (2005)
21. McCormick, F.: Future prospects for oncolytic therapy. Oncogene **24**(52), 7817–9 (2005)
22. Russell, S.J., Peng, K.W., Bell, J.C.: Oncolytic virotherapy. Nat. Biotechnol. **30**(7), 658–670 (2012)
23. Lorence, R.M., Pecora, A.L., Major, P.P., Hotte, S.J., Laurie, S.A., Roberts, M.S., Groene, W.S., Bamat, M.K.: Overview of phase I studies of intravenous administration of PV701, an oncolytic virus. Curr. Opin. Mol. Ther. **5**(6), 618–624 (2003)
24. Roberts, M.S., Lorence, R.M., Groene, W.S., Bamat, M.K.: Naturally oncolytic viruses. Curr. Opin. Mol. Ther. **8**(4), 314–321 (2006)
25. De Pillis, L., Mallet, D., Radunskaya, A.: Spatial tumor-immune modeling. Comput. Math. Methods Med. **7**(2–3), 159–176 (2006)
26. Gerlee, P., Anderson, A.: An evolutionary hybrid cellular automaton model of solid tumour growth. J. Theor. Biol. **246**(4), 583–603 (2007)
27. Deisboeck, T., Zhang, L., Yoon, J., Costa, J.: In silico cancer modeling: is it ready for prime time? Nat. Clin. Pract. Oncol. **6**(1), 34–42 (2008)
28. Anderson, A., Quaranta, V.: Integrative mathematical oncology. Nat. Rev. Cancer **8**(3), 227–234 (2008)
29. Chaplain, M., McDougall, S., Anderson, A.: Mathematical modeling of tumor-induced angiogenesis. Annu. Rev. Biomed. Eng. **8**, 233–257 (2006)
30. Anderson, A., Chaplain, M.: Continuous and discrete mathematical models of tumor-induced angiogenesis. Bull. Math. Biol. **60**(5), 857–899 (1998)
31. Byrne, H., Chaplain, M.: Growth of necrotic tumors in the presence and absence of inhibitors. Math. Biosci. **135**(2), 187–216 (1996)
32. Macklin, P., McDougall, S., Anderson, A., Chaplain, M., Cristini, V., Lowengrub, J.: Multiscale modelling and nonlinear simulation of vascular tumour growth. J. Math. Biol. **58**(4), 765–798 (2009)
33. Lowengrub, J., Frieboes, H., Jin, F., Chuang, Y., Li, X., Macklin, P., Wise, S., Cristini, V.: Nonlinear modelling of cancer: bridging the gap between cells and tumours. Nonlinearity **23**(1), R1 (2009)
34. Swierniak, A., Ledzewicz, U., Schattler, H.: Optimal control for a class of compartmental models in cancer chemotherapy. Int. J. Appl. Math. Comput. Sci. **13**(3), 357–368 (2003)
35. Ledzewicz, U., Schättler, H.: Analysis of a cell-cycle specific model for cancer chemotherapy. J. Biol. Syst. **10**(03), 183–206 (2002)
36. Kim, P., Lee, P., Levy, D.: Dynamics and potential impact of the immune response to chronic myelogenous leukemia. PLoS Comput. Biol. **4**(6), e1000095 (2008)
37. Hinow, P., Gerlee, P., McCawley, L., Quaranta, V., Ciobanu, M., Wang, S., Graham, J., Ayati, B., Claridge, J., Swanson, K., et al.: A spatial model of tumor-host interaction: application of chemotherapy. Math. Biosci. Eng **6**(3), 521 (2009)

38. Hinow, P., Rogers, C., Barbieri, C., Pietenpol, J., Kenworthy, A., DiBenedetto, E.: The DNA binding activity of p53 displays reaction-diffusion kinetics. Biophys. J. **91**(1), 330–342 (2006)
39. Enderling, H., Chaplain, M., Anderson, A., Vaidya, J.: A mathematical model of breast cancer development, local treatment and recurrence. J. Theor. Biol. **246**(2), 245–259 (2007)
40. Enderling, H., Anderson, A., Chaplain, M., Rowe, G., et al.: Visualisation of the numerical solution of partial differential equation systems in three space dimensions and its importance for mathematical models in biology. Math. Biosci. Eng. **3**(4), 571 (2006)
41. Deisboeck, T., Berens, M., Kansal, A., Torquato, S., Stemmer-Rachamimov, A., Chiocca, E.: Pattern of self-organization in tumour systems: complex growth dynamics in a novel brain tumour spheroid model. Cell Prolif. **34**(2), 115–134 (2008)
42. Hanin, L.: Identification problem for stochastic models with application to carcinogenesis, cancer detection and radiation biology. Discrete Dyn. Nat. Soc. **7**(3), 177–189 (2002)
43. Bartoszyński, R., Edler, L., Hanin, L., Kopp-Schneider, A., Pavlova, L., Tsodikov, A., Zorin, A., Yakovlev, A.: Modeling cancer detection: tumor size as a source of information on unobservable stages of carcinogenesis. Math. Biosci. **171**(2), 113–142 (2001)
44. Swierniak, A., Kimmel, M., Smieja, J.: Mathematical modeling as a tool for planning anticancer therapy. Eur. J. Pharmacol. **625**(1), 108–121 (2009)
45. Marciniak-Czochra, A., Kimmel, M.: Modelling of early lung cancer progression: influence of growth factor production and cooperation between partially transformed cells. Math. Models Methods Appl. Sci. **17**(supp 01), 1693–1719 (2007)
46. Shochat, E., Hart, D., Agur, Z.: Using computer simulations for evaluating the efficacy of breast cancer chemotherapy protocols. Math. Models Methods Appl. Sci. **9**(04), 599–615 (1999)
47. Gaffney, E.: The application of mathematical modelling to aspects of adjuvant chemotherapy scheduling. J. Math. Biol. **48**(4), 375–422 (2004)
48. Gatenby, R., Maini, P.: Mathematical oncology: cancer summed up. Nature **421**(6921), 321–321 (2003)
49. Bellomo, N., Li, N., Maini, P.: On the foundations of cancer modelling: selected topics, speculations, and perspectives. Math. Models Methods Appl. Sci. **18**(04), 593–646 (2008)
50. Cristini, V., Lowengrub, J.: Multiscale Modeling of Cancer: An Integrated Experimental and Mathematical Modeling Approach. Cambridge University Press, Cambridge (2010)

Part I
Treatment of cancer with small molecule inhibitors

Chapter 2
An Introduction to Small Molecule Inhibitors and Chronic Myeloid Leukemia

Abstract The first part of the book is concerned with the targeted treatment of cancers with small molecule inhibitors. These are inhibitors that have been designed to counter specific cellular defects that are responsible for initiating and maintaining the disease. The best studied example so far is the treatment of chronic myeloild leukemia (CML) with tyrosine kinase inhibitors, such as imatinib, which has led to impressive treatment responses. Yet, drug-resistant tumor cells present an important barrier to successful treatment, especially in the advanced phase of the disease, and can lead to treatment failure. The dynamics of CML and the evolution of drug-resistant mutants will be investigated in detail in this part of the book, through the lens of mathematical models. The current chapter provides the necessary biological background to the mathematical models.

Keywords Chronic myeloid leukemia (CML) · Chronic phase · Accelerated phase · Blast crisis · Symptoms · Tyrosine kinase inhibitors · BCR-ABL · Imatinib

2.1 Basics

One of the greatest challenges for cancer therapy is to effectively kill cancer cells while leaving healthy tissue largely unharmed. The most widespread treatment approaches include chemotherapy and radiation therapy. These treatments damage dividing cells, and thus can lead to the death of cancer cells. However, healthy cells that divide are also hit, leading to strong side effects. In other words, these treatments are not specific to cancer cells. Because chemo- and radiation treatments cause genetic alterations in cells, they may not only induce side effects, but can also induce the formation of new tumors. To make things even more difficult, tumor cells can easily acquire resistance to these treatments, including simultaneous resistance to multiple therapeutic agents. Thus, there is a pressing need to develop new treatments that show both a high effectiveness and a high specificity for cancer cells.

N. L. Komarova and D. Wodarz, *Targeted Cancer Treatment in Silico*,
Modeling and Simulation in Science, Engineering and Technology,
DOI: 10.1007/978-1-4614-8301-4_2, © Springer Science+Business Media New York 2014

One of such approaches is small molecule inhibitors [1, 2] that target specific defects in cancer cells, leading to their death. This type of approach is based on our understanding of the cellular processes that become corrupted in cancer cells and that induce uncontrolled growth, allowing us to disrupt this mechanism. The treatment of cancer with small molecule inhibitors is the subject of the first part of the book. We will investigate the tumor dynamics during therapy, examine the evolutionary dynamics of drug resistant mutants, and discuss treatment strategies aimed at preventing treatment failure as a result of drug resistance. The best and most successful example of treatment with small molecule inhibitors so far is the treatment of chronic myeloid leukemia (CML) with tyrosine kinase inhibitors [2–5]. Initially, patients were treated with imatinib mesylate (Gleevec) [6, 7]. Subsequently, second-generation inhibitors have been developed, such as dasatinib and nilotinib [8–10]. Further agents are under development [4, 11]. The small molecule inhibitors listed above are currently used to treat CML. Apart from CML, there are other cancers that respond to various types of small molecule inhibitors, see e.g., [1], which lists 11 kinase inhibitors that have received US Food and Drug Administration approval as cancer treatments. For example, gefitinib is used to treat non-small cell lung cancer [12]. Erlotinib is used to treat non-small cell lung cancer [13] and some pancreatic cancers [14]. Bortezomib has been approved for treatment of multiple myeloma [15]. Sorafenib is used for treatment of advanced renal-cell carcinomas [16] and hepatocellular carcinomas [17].

In this part of the book, we will discuss cancer treatments with small molecule inhibitors. Because the oldest and the best studied cancer-drug system in the context of small molecule inhibitors is CML and the drug imatinib, and because the kind of modeling we use in this part of the book is best justified for non-solid tumors, we will illustrate all our findings by using the example of CML. The current chapter provides the basic biological background on CML that is required to understand the modeling. Explaining deeper biological details would go beyond the scope of this book, and this information is found in many papers on the topic, some of which are cited in this chapter.

2.2 Natural History of CML

CML is a cancer of the hematopoietic system, i.e., a cancer of the blood (leukemia) [4, 18, 19]. It is characterized by an unregulated growth of myeloid cells in the bone marrow, and an accumulation of those cells in the blood. CML is not the most common leukemia, accounting only for about 20 % of all leukemia cases. According to the Surveillance Epidemiology and End Results (SEER) statistics, from 2005–2009, the median age at diagnosis for CML was 64 years of age [20]. The disease process can be divided into three phases: the chronic phase, the accelerated phase, and the blast crisis. The disease starts with the chronic phase which tends to be asymptomatic. The number of cells grows relatively slowly leading to an elevated blood cell count, and the cells are characterized by a relatively high degree of differentiation. The chronic phase eventually develops into the accelerated phase, where the cell population expands more rapidly and the proportion of undifferentiated cells (blasts) increases. Finally, the blast crisis is characterized by the explosive growth of blast cells that show low degrees of differentiation. It presents a growth pattern similar to acute leukemia, leading to severe pathology and death. While CML was first described in the late nineteenth century, the defining characteristic of this tumor was discovered in 1960: 90 % of affected individuals carry the Philadelphia chromosome, which is the result of a fusion between abelson (ABL) tyrosine kinase gene on chromome 9 and the break point cluster (BCR) gene on chromosome 22. The BCR-ABL fusion gene is an oncogene that is thought to maintain and drive the disease.

2.3 Therapy

The aim of therapy is to reduce the disease burden. In this respect, three types of therapy responses have been considered [4]: the hematological response, the cytogenetic response, and the molecular response. A complete hematological response is defined by the normalization of the blood cell counts. The cytogenetic response is the most common assessment of treatment, quantifying the number of Philadelphia chromosome positive metaphases over time. An absence of measurable levels of these cells is called a complete cytogenetic response, while a major cytogenetic response is defined by a prevalence of Philadelphia chromosome positive cells of about 0–35 %. The molecular response quantifies the number of BCR-ABL mRNAs and is the most sensitive method to monitor the treatment response. A complete molecular response is defined by the absence of detectable BCR-ABL mRNAs, measured by reverse transcriptase polymerase chain reaction (RT-PCR). A major molecular response occurs if a 3-log reduction of the BCR-ABL/BCR level occurs compared to the median pre-treatment levels.

The initial treatment approach involved the use of chemotherapeutic agents [21] which normalized the blood cell count but failed to alter disease progression to blast crisis. Treatment outcomes were significantly improved by the introduction of recombinant interferon alpha (rIFN-α) therapy [22]. Hematopoietic stem cell transplants also offered benefits and could even lead to cures, although they are associated with a high risk of morbidity and mortality [4]. The breakthrough in CML treatment came with the development of small molecule inhibitors of the tumorigenic BCR-ABL tyrosine kinase. The first such small molecule inhibitor was imatinib mesylate, also referred to as imatinib, STI-571, or Gleevec [6, 7]. The principle according to which this drug works is as follows. The BCR-ABL gene product is a constitutively active kinase that binds ATP and transfers phosphate from ATP to tyrosine residues, thereby leading to aberrant cellular behavior. Imatinib is designed to specifically block the binding of ATP to the BCR-ABL tyrosine kinase, thereby preventing continued functioning of the tumor cells (Fig. 2.1).

The use of imatinib in the chronic phase of the disease led to very good treatment results, demonstrating impressive hematological, cytogenetic, and molecular responses. Patients with complete cytogenetic and molecular responses also showed

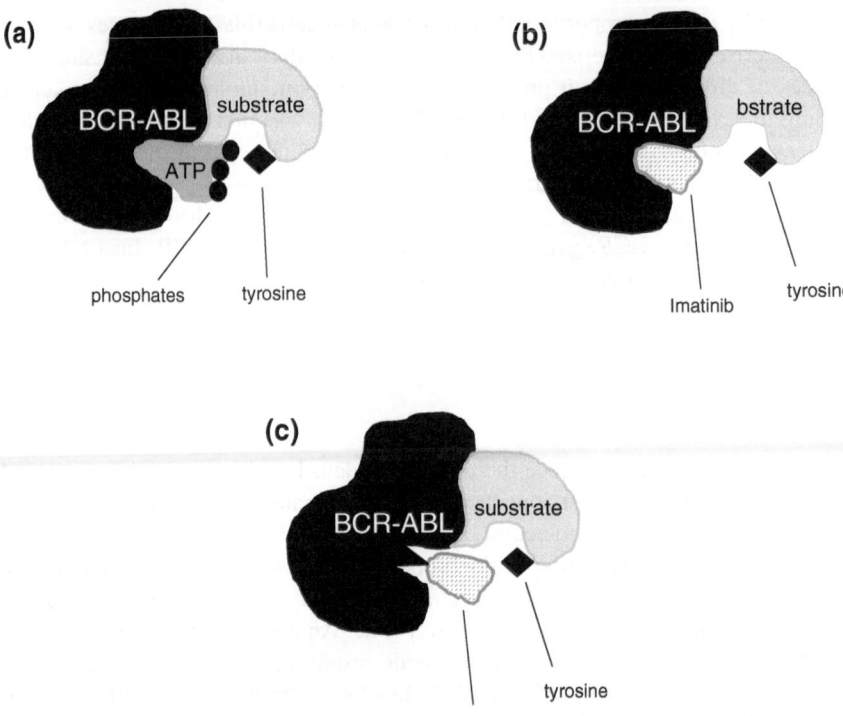

Fig. 2.1 Schematic explaining the concept behind the tyrosine kinase inhibitor imatinib in the treatment of CML. (**a**) BCR-ABL tyrosine kinase binds ATP and transfers phosphates from ATP to tyrosine residues on substrates. This leads to downstream effects that cause unregulated cellular behavior. (**b**) Tyrosine kinase inhibitors, such as imatinib, block ATP from binding to BCR-ABL tyrosine kinase, thus preventing unregulated behavior. (**c**) A point mutation in the ATP binding site of the BCR-ABL tyrosine kinase changes the shape of the binding site (indicated by a *triangle*), thus blocking the drug from binding

prolonged progression-free survival, indicating that the course of the disease had been altered [4]. However, in patients that were treated during blast crisis, the outcome was less impressive. A major obstacle for treatment success was resistance to imatinib [23–28]. Most commonly, resistance is conferred by point mutations that prevent the binding of the drug to its target (Fig. 2.1). In a minority of cases, gene amplification events can also lead to drug resistance. The consequent increase in the level of BCR-ABL protein levels allows ongoing oncogenic activity in the presence of the drug. Subsequent, second-generation inhibitors have been developed to broaden the therapeutic approach and to address problems arising from drug resistance. These include the drugs dasatinib and nilotinib [8–10].

There are are many point mutations that confer resistance against a particular inhibitor. Many of them specifically confer resistance to one (e.g., imatinib) but not the other inhibitors (e.g., dasatinib and nilotinib). In other words, they do not show cross-resistance. However, one of the most common mutations found in blast crisis patients, the T315I mutation [29], does show cross-resistance and confers resistance to imatinib, dasatinib, and nilotinib. This can considerably complicate treatment options. A number of drugs are under development that are aimed at overcoming the T315I mutation [4, 30]. Examples of such drugs include ponatinib, a pan-BCR-ABL inhibitor, and the aurora kinase inhibitor MK-0457, which have documented activity against cells containing the T315I mutation.

2.4 The Role of Mathematical Models

In addition to clinical and experimental research, mathematical and computational models can be very useful to study the treatment of CML with tyrosine kinase inhibitors. Mathematical models, in combination with data, can shed light on the mechanisms that underlie the dynamics of BCR-ABL transciript decline during treatment. Moreover, mathematical models allow us to investigate the evolutionary dynamics of drug-resistant strains, which in turn can lead to the design of treatment strategies aimed at overcoming treatment failure due to resistance. Based on defined assumptions, the models provide insight and make predictions that would be impossible to achieve by experimental and clinical study alone.

One of the problems with modeling tumor growth, progression, and therapy is the enormous complexity that underlies the process of carcinogenesis, and our lack of understanding of the corresponding processes. In many cancers, a sequence of multiple mutations has to be accumulated by cells in order for them to undergo uncontrolled proliferation, and the exact pathways to cancer are often poorly understood. With CML, it appears that a single oncogene, the BCR-ABL fusion gene, is responsible for initiating and maintaining the disease. Without BCR-ABL activity, the cells cannot divide out of control, and die. This represents a relatively simple setting, which in turn reduces the uncertainty that arises when formulating biological processes in terms of mathematical equations. In addition, the phase of the disease that is most complicated to treat and where resistance poses the biggest threat is the blast crisis, which appears to be characterized by explosive, exponential growth of the cells [31]. This is again easily described by mathematical models and avoids complications and uncertainties regarding the formulation of more complex tumor growth patterns that are found in many cancers and that are harder to understand [32].

The following chapters will describe how mathematical models have been used to understand basic patterns of CML treatment dynamics as well as the evolution of drug resistant cells. First, we concentrate on basic dynamics, examining the patterns of BCR-ABL transcript number decline during treatment. Then, the evolutionary dynamics of drug-resistant mutants will be considered. This will be done on various levels of complexity. We will start with the simplest scenario where a single, exponentially increasing, population of tumor cells is considered and where we only concentrate on drug resistance mutations that are not characterized by cross-resistance. Subsequently, we will introduce further biological complexity and investigate how this influences the results. Thus, a tumor stem cell compartment will be considered explicitly, and the effect of the T315I cross-resistant mutations will be investigated. Special emphasis will be placed on treatment strategies that are designed to avoid resistance-induced treatment failure. All this analysis requires a set of mathematical tools. Thus, before exploring the evolutionary models of drug resistance, a chapter (Chap. 4) will be devoted to the general mathematical methodology that underlies this approach, which readers with a biological background may skip. In addition, some of the subsequent chapters contain a Methodology section, which describes further mathematical tools that are particular to the derivations of results described in the corresponding chapter. Again, these sections can be skipped by a biological readership. The description of the results does not depend on the knowledge of these techniques.

References

1. Zhang, J., Yang, P., Gray, N.: Targeting cancer with small molecule kinase inhibitors. Nat. Rev. Cancer **9**(1), 28–39 (2009)
2. Druker, B.J.: Imatinib as a paradigm of targeted therapies. Adv. Cancer Res. **91**, 1–30 (2004)
3. Deininger, M.W., Goldman, J.M., Melo, J.V.: The molecular biology of chronic myeloid leukemia. Blood **96**(10), 3343–3356 (2000)
4. An, X., Tiwari, A., Sun, Y., Ding, P., Ashby, C., Chen, Z.: Bcr-abl tyrosine kinase inhibitors in the treatment of philadelphia chromosome positive chronic myeloid leukemia: a review. Leuk. Res. **34**(10), 1255–1268 (2010)
5. Ferdinand, R., Mitchell, S., Batson, S., Tumur, I.: Treatments for chronic myeloid leukemia: a qualitative systematic review. J. Blood Med. **3**, 51 (2012)
6. Deininger, M., Buchdunger, E., Druker, B.: The development of imatinib as a therapeutic agent for chronic myeloid leukemia. Blood **105**(7), 2640–2653 (2005)
7. Deininger, M.W., Druker, B.J.: Specific targeted therapy of chronic myelogenous leukemia with imatinib. Pharmacol. Rev. **55**(3), 401–423 (2003)
8. Leitner, A., Hochhaus, A., Müller, M., et al.: Current treatment concepts of cml. Curr. Cancer Drug Targets **11**(1), 31 (2011)
9. Kantarjian, H., Baccarani, M., Jabbour, E., Saglio, G., Cortes, J.: Second-generation tyrosine kinase inhibitors: the future of frontline cml therapy. Clin. Cancer Res. **17**(7), 1674–1683 (2011)
10. Weisberg, E., Manley, P., Cowan-Jacob, S., Hochhaus, A., Griffin, J.: Second generation inhibitors of bcr-abl for the treatment of imatinib-resistant chronic myeloid leukaemia. Nat. Rev. Cancer **7**(5), 345–356 (2007)
11. Karvela, M., Helgason, G., Holyoake, T.: Mechanisms and novel approaches in overriding tyrosine kinase inhibitor resistance in chronic myeloid leukemia. Expert Rev. Anticancer Ther. **12**(3), 381–392 (2012)
12. Paez, J.G., Janne, P.A., Lee, J.C., Tracy, S., Greulich, H., Gabriel, S., Herman, P., Kaye, F.J., Lindeman, N., Boggon, T.J., et al.: Egfr mutations in lung cancer: correlation with clinical response to gefitinib therapy. Sci. Signal. **304**(5676), 1497 (2004)
13. Pao, W., Miller, V., Zakowski, M., Doherty, J., Politi, K., Sarkaria, I., Singh, B., Heelan, R., Rusch, V., Fulton, L., et al.: Egf receptor gene mutations are common in lung cancers from never smokers and are associated with sensitivity of tumors to gefitinib and erlotinib. Proc. Natl. Acad. Sci. U S A **101**(36), 13306–13311 (2004)

14. Miyabayashi, K., Ijichi, H., Mohri, D., Tada, M., Yamamoto, K., Asaoka, Y., Ikenoue, T., Tateishi, K., Isayama, H., et al.: Erlotinib prolongs survival in pancreatic cancer by blocking gemcitabine-induced mapk signals. Cancer Res. **73**(7), 2221–2234 (2013)
15. Adams, J., Kauffman, M.: Development of the proteasome inhibitor velcade (bortezomib). Cancer Invest. **22**(2), 304–311 (2004)
16. Escudier, B., Eisen, T., Stadler, W.M., Szczylik, C., Oudard, S., Siebels, M., Negrier, S., Chevreau, C., Solska, E., Desai, A.A., et al.: Sorafenib in advanced clear-cell renal-cell carcinoma. N. Engl. J. Med. **356**(2), 125–134 (2007)
17. Llovet, J.M., Ricci, S., Mazzaferro, V., Hilgard, P., Gane, E., Blanc, J.F., de Oliveira, A.C., Santoro, A., Raoul, J.L., Forner, A., et al.: Sorafenib in advanced hepatocellular carcinoma. N. Engl. J. Med. **359**(4), 378–390 (2008)
18. Melo, J.V., Hughes, T.P., Apperley, J.F.: Chronic myeloid leukemia. Hematology: Am Soc Hematol Educ Book, **2003**(1) pp. 132–52 (2003).
19. Melo, J.V., Barnes, D.J.: Chronic myeloid leukaemia as a model of disease evolution in human cancer. Nat. Rev. Cancer **7**(6), 441–453 (2007)
20. Howlader, N., Noone, A., Krapcho, M., Neyman, N., Aminou, R., Altekreuse, S., Kosary, C., Ruhl, J., Tatalovich, Z., Cho, H., et al.: Seer Cancer Statistics Review, 1975–2009 (Vintage 2009 Populations). National Cancer Institute, Bethesda (2012)
21. Bolin, R., Robinson, W., Sutherland, J., Hamman, R.: Busulfan versus hydroxyurea in long-term therapy of chronic myelogenous leukemia. Cancer **50**(9), 1683–1686 (1982)
22. Group, C.M.L.T.C.: Interferon alfa versus chemotherapy for chronic myeloid leukemia: a meta-analysis of seven randomized trials. J. Natl. Cancer Inst. **89**, 1616–1620 (1997)
23. Shannon, K.M.: Resistance in the land of molecular cancer therapeutics. Cancer Cell **2**(2), 99–102 (2002)
24. Shah, N.P., Tran, C., Lee, F.Y., Chen, P., Norris, D., Sawyers, C.L.: Overriding imatinib resistance with a novel abl kinase inhibitor. Science **305**(5682), 399–401 (2004)
25. Tauchi, T., Ohyashiki, K.: Molecular mechanisms of resistance of leukemia to imatinib mesylate. Leuk. Res. **28**(Suppl 1), S39–45 (2004)
26. Kantarjian, H., Talpaz, M., Giles, F., O'Brien, S., Cortes, J., et al.: New insights into the pathophysiology of chronic myeloid leukemia and imatinib resistance. Ann. Intern. Med. **145**(12), 913 (2006)
27. Branford, S., Rudzki, Z., Walsh, S., Parkinson, I., Grigg, A., Szer, J., Taylor, K., Herrmann, R., Seymour, J., Arthur, C., et al.: Detection of bcr-abl mutations in patients with cml treated with imatinib is virtually always accompanied by clinical resistance, and mutations in the atp phosphate-binding loop (p-loop) are associated with a poor prognosis. Blood **102**(1), 276–283 (2003)
28. Volpe, G., Panuzzo, C., Ulisciani, S., Cilloni, D., et al.: Imatinib resistance in cml. Cancer Lett. **274**(1), 1 (2009)
29. O'Hare, T., Corbin, A., Druker, B.: Targeted cml therapy: controlling drug resistance, seeking cure. Curr. Opin. Genetics Dev. **16**(1), 92–99 (2006)
30. Giles, F., Cortes, J., Jones, D., Bergstrom, D., Kantarjian, H., Freedman, S.: Mk-0457, a novel kinase inhibitor, is active in patients with chronic myeloid leukemia or acute lymphocytic leukemia with the t315i bcr-abl mutation. Blood **109**(2), 500–502 (2007)
31. Calabretta, B., Perrotti, D.: The biology of cml blast crisis. Blood **103**(11), 4010–4022 (2004)
32. Rodriguez-Brenes, I.A., Komarova, N.L., Wodarz, D.: Evolutionary dynamics of feedback escape and the development of stem-cell-driven cancers. Proc. Natl. Acad. Sci. U S A **108**(47), 18983–18988 (2011)

Chapter 3
Basic Dynamics of Chronic Myeloid Leukemia During Imatinib Treatment

Abstract Treatment with tyrosine kinase inhibitors, such as imatinib, have led to impressive therapy responses in the clinic. This chapter will discuss the pattern of decline of the BCR-ABL transcript numbers during treatment, and explore different hypotheses to explain them. Treatment typically results in a bi-phasic decline of BCR-ABL transcript numbers, where a faster phase of decline is followed by a slower phase. The earliest hypotheses tried to explain this pattern by assuming that more differentiated tumor cells are susceptible to the drug, while tumor stem cells are resistant. Subsequent work showed that the bi-phasic decline, and other, less common decline patterns found among patients, can be explained by differential susceptibility to the drug by activated and quiescent tumor stem cells. In addition to the basic tumor cell dynamics, mathematical models have also indicated that a transient rise of anti-tumor immunity during treatment could contribute to determining the pattern of treatment response. Implications of the different hypotheses for treatment strategies are discussed.

Keywords CML · BCR-ABL · Treatment · Bi-phasic decline · Stem cells · Cellular quiescence · Cycling cells · Tumor heterogeneity · Immunity · Resistance · T cells · CD8+ cells · CD4+ cells · Population dynamics of cells · Long-term remission

3.1 Introduction

As discussed in the previous chapter, the quantification of the BCR-ABL transcript numbers allows tracking of the CML tumor burden during tumor growth and treatment. While the number of BCR-ABL transcripts does not equal the number of CML tumor cells, they do reflect the dynamics of the cancer cells over time. This gives rise to the possibility to quantify the population dynamics of CML cells during targeted therapy. In the context of imatinib (Gleevec) treatment, this has led to the emergence of the following patterns [1, 2]. On average, a bi-phasic decline of BCR-ABL

N. L. Komarova and D. Wodarz, *Targeted Cancer Treatment in Silico*,
Modeling and Simulation in Science, Engineering and Technology,
DOI: 10.1007/978-1-4614-8301-4_3, © Springer Science+Business Media New York 2014

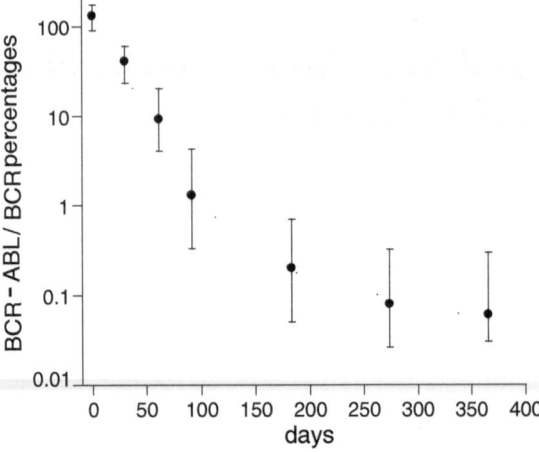

Fig. 3.1 Average dynamics of CML during imatinib treatment, taken from [1]. The BCR-ABL/BCR ratios are plotted over time, averaged over many patients. Patients that relapsed during therapy were excluded. A bi-phasic decline is observed, where an initial fast decline phase is followed by a slower decline phase. This was the first study that published such averaged time series, and a similar average dynamics were found in patients from the German cohort of the IRIS study [2], not shown here

transcript numbers is observed (Fig. 3.1), with an initial fast phase of decline followed by a second slower phase. In the case of treatment failure, BCR-ABL transcript numbers subsequently rose again. The reason for this bi-phasic decline remains unclear. Several theories have been developed, some of which will be discussed in this chapter and evaluated in the context of the observed tumor cell dynamics during therapy. Likely more than one mechanism is responsible for explaining the clinical data. Insights into the basic tumor cell dynamics during treatment also provides an important foundation for understanding the evolutionary dynamics of CML, particularly with respect to the generation of drug-resistant mutants, which will be explored in subsequent chapters.

3.2 Basic Hypothesis: Differentiated Versus Stem Cells

A prevalent hypothesis in the field of cancer research is that the cellular architecture of tumors is related to that found in healthy tissue [3]. Healthy tissue is maintained by a small number of tissue stem cells. Division of the tissue stem cells leads either to the generation of more stem cells (self-renewal), or to the generation of transit-amplifying cells that undergo a certain number of cell divisions before becoming terminally differentiated cells. Terminally differentiated cells do not divide anymore and die after a given period of time. Similarly, tumors are thought to consist of a relatively small population of "tumor stem cells," while the bulk of the tumor is made up of "differentiated tumor cells." It is thought that therapies often impact the bulk of differentiated tumor cells while not affecting the tumor stem cell population. Hence, the tumor stem cells persist during treatment and thus maintain the cancer. In this spirit, it has been suggested that the bi-phasic decline of BCR-ABL transcripts during imatinib treatment can be explained by the differential susceptibility of tumor stem cells and more differentiated cells [1]. While the more differentiated cells are susceptible to imatinib-mediated activity, the tumor stem cells are resistant to therapy according to this hypothesis. When formulated in terms of a mathematical model, this hypothesis can account for the observed bi-phasic decay of BCR-ABL transcripts during Imatinib treatment.

3.3 Active Versus Quiescent Tumor Stem Cells

Stem cells can be in an active, dividing, state and in a quiescent state. The same has been found to apply to cancer stem cells. While some have argued that CML stem cells are not susceptible to targeted treatment with imatinib (see above), others have argued that the susceptibility of CML stem cell to imatinib treatment depends on the activation status of the cancer stem cells. While quiescent CML stem cells are resistant to treatment, active CML stem cells are susceptible. This has important implications for understanding the decline dynamics of BCR-ABL transcript numbers during imatinib therapy, and is explored with mathematical models as follows.

We consider an ordinary differential equation model that tracks the average number of proliferating and quiescent cells, $x(t)$ and $y(t)$, as a function of time. The proliferating cells divide with a rate L and die with a rate d. The death rate may capture both the natural death rate of cancer cells, D, and the treatment-induced death rate, H. In the absence of treatment, we have $d = D$ and $L > d$, that is, the cell population grows exponentially. Treatment increases the parameter $d = D + H$. If treatment is efficient, then $L < d$, such that the tumor cell population declines. The cells enter a quiescent state with a rate α, and quiescent cells re-enter the cell cycle with a rate β. Note that quiescent cells do not divide or die and are not susceptible to any drug activity. The model is given by the following set of equations:

$$\dot{x} = (L - d - \alpha)x + \beta y, \tag{3.1}$$
$$\dot{y} = \alpha x - \beta y, \tag{3.2}$$

Note that this model does not explicitly take into account differentiated CML cells. These are not thought to contribute significantly to malignant growth and are simply proportional to the number of primitive CML cells.

Assume the existence of a number of primitive CML cells, a fraction of which is quiescent. They are treated with the drug imatinib. In this first model, we assume that all CML cells are susceptible to the drug and that no drug resistance is generated by mutation. In this scenario, the death rate of the CML cells is greater than their division

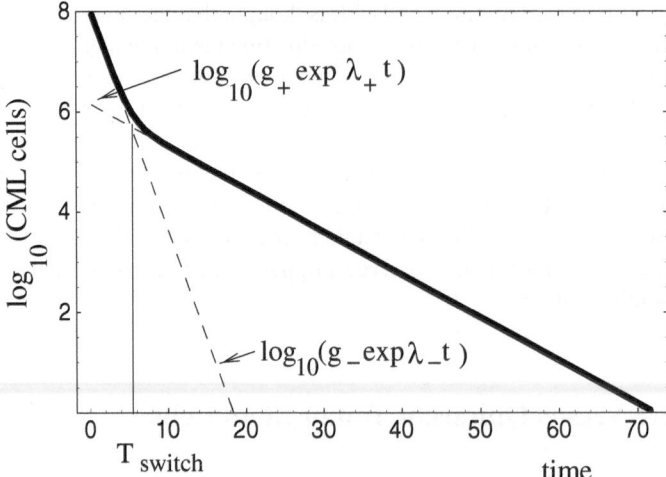

Fig. 3.2 Predicted bi-phasic decline of the CML cell population as a function of time, for parameters $l = 1, d = 1.5, \alpha = 0.01, \beta = 0.2, I_0 = 10^8$ and $J_0 = 10^2$. The solid line represents $\log_{10}(x(t) + y(t))$, and the dashed lines are $\log_{10}(g_+\exp\{\lambda_+t\})$ and $\log_{10}(g_-\exp\{\lambda_-t\})$, where g_+ and g_- are constants calculated by standard methods from the linear Eqs. (3.1 and 3.2). The time of treatment in this case is $T_{\text{treat}} = 72.1$ and time of switching between the two phases is $T_{\text{switch}} = 5.1$

rate $(d > L)$, such that the population of cells declines. The model suggests various behaviors upon initiation of treatment. In one parameter region, therapy results in two distinct phases of exponential decline (Fig. 3.2), as observed in experimental data [1, 2]. First, the population of cells declines exponentially with a relatively fast rate, λ_-, as a result of the death of proliferating cells, x. Then, a slower phase of exponential decline at a rate λ_+ is observed because the quiescent cells become dominant and are only killed when they wake up and re-enter a cycling state. The values $\lambda_\pm$ are given by

$$\lambda_\pm = \frac{1}{2}\left(d - L + \alpha + \beta \pm \sqrt{(d - L + \alpha + \beta)^2 - 4\beta(d - L)}\right)$$

(see e.g., [4] for solution methods of ODEs). For small values of α the expressions for the decay rates simplify and we have $\lambda_+ = -(d - L)$ and $\lambda_- = -\beta$, that is, the first wave of decline happens at the net decay rate of cycling cells and the second wave happens at the rate of cell awakening.

In this model, treatment will eventually drive the tumor to extinction, but the time it takes to achieve this goal is influenced by the kinetics of the second, slower phase of decline, and thus by the rate at which cells enter the quiescent state, and the rate at which cells exit the quiescent state. The higher the rate at which cells enter quiescence, and the slower the rate at which cells exit quiescence, the longer it takes to reduce the CML population towards extinction. Also, the lower the overall death rate of cells, the longer it takes to reduce the tumor towards extinction. In the

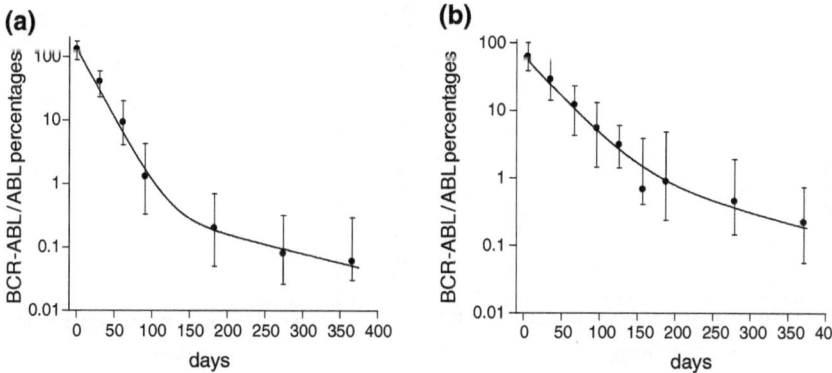

Fig. 3.3 The relative amount of CML cells as a function of time, in patients treated with imatinib. The *circles* represent experimental data replotted from (**a**) [1] and (**b**) [2]; they show the median values of BCR-ABL transcripts (relative to BCR transcripts in (**a**) and ABL transcripts in (**b**)). The vertical bars are the quartiles. The solid lines represent the fitted theoretical curves, formula (7) of Text S1, obtained by a mean-square procedure. The estimated parameter values are: (**a**) $d - L = 0.0502$ days^{-1}, $\beta = 0.0065$ days^{-1}, $\alpha = 10^{-5}$ days^{-1}, $J_0 = 0.47$; (**b**) $d - L = 0.0278$ days^{-1}, $\beta = 0.0067$ days^{-1}, $\alpha = 0.0004$ days^{-1}, $J_0 = 0.50$. Here J_0 denotes the initial percentage of quiescent CML cells

model, the time of the switch between the two phases of decline (T_{switch}, see Fig. 3.2) is proportional to $\frac{1}{|\lambda_+ - \lambda_-|}$, and the time of extinction is proportional to $\frac{1}{|\lambda_+|}$.

Note, however, that these dynamics are not universal in the model. This type of biphasic decline occurs if the death rate of cells is larger than the sum of the division and quiescence rates ($d > L + \alpha + \beta$). For smaller death rates, when this condition is not fulfilled, two further patterns of decline are observed. Either the population of cells declines in a single exponential phase during treatment, or a first and slower phase of cell decline is followed by a second and faster phase of cell decline (a reverse biphasic decline). This behavior is observed if there is more quiescence in the population of tumor cells. In this case, the first phase need not be the fastest anymore, because it can be dictated by the kinetics of cell activation rather than cell death. Once a sufficient number of cells has been activated, cell death is the dominant factor and the rate of cell decline speeds up.

In order to show that our equations can accurately describe clinical data, the model was fitted to two data sets that document a bi-phasic decline of CML cells during treatment (Fig. 3.3). The first data set is taken from [1] and contains median BCR-ABL transcript levels from a selected cohort ($n = 68$) that excludes cases with transiently increasing BCR-ABL transcript levels (Fig. 3.3a). The second data set is taken from [2] and contains median BCR-ABL transcript levels from an unselected cohort ($n = 69$) of CML patients (Fig. 3.3b). In addition to the median values, Roeder et al presented individual responses to imatinib therapy. Figure 3.4 re-plots the clinical data from two patients that do not show a bi-phasic decline. Based on our model, it can be hypothesized that in these patients the number of CML cells

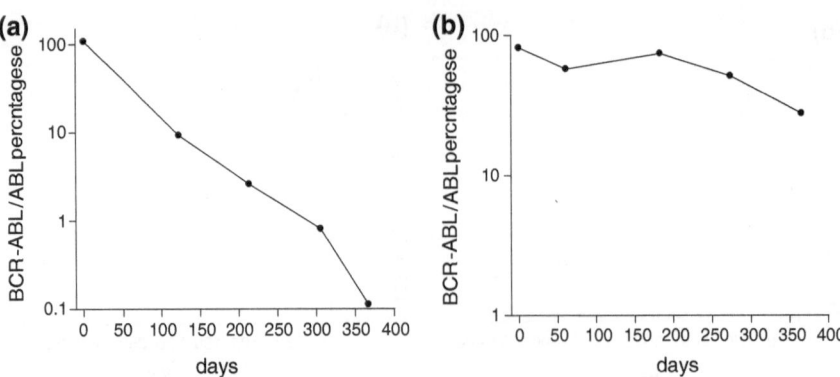

Fig. 3.4 Data that document the decline of CML cells during imatinib treatment in two patients, taken from [2]

declines in a single exponential phase during treatment (Fig. 3.4a), or according to the reverse biphasic decline pattern (Fig. 3.4b). However, analysis of additional data for longer periods of time will be necessary to test this hypothesis.

In summary, whether or not CML can be cured by imatinib therapy in the absence of acquired resistance depends on the time it takes for the cancer cells to be driven extinct by the treatment, and this in turn depends on the rate constants. Eventual CML extinction is the only theoretically possible outcome in the presence of therapy, but it may be achieved after a period of time that is longer than the life-span of the patient. Variations in parameters that determine the kinetics of cellular quiescence can determine whether relapse is observed in patients that stop imatinib treatment after a certain amount of time [5]. Note that our notion of treatment induced "cancer extinction" is a mathematical one, that is, in the model we analyze here, the cancer cell population goes extinct, which corresponds to a cure. In patients, however, other complicating factors not included in this model may render tumor extinction a difficult goal to achieve by treatment. Therefore, our mathematical notion of "tumor extinction" should be translated into "clinical remission" in a medical context.

3.4 Heterogeneity and Correlations in the Bi-Phasic Decline of BCR-ABL Transcript Levels

At the next level of complexity, rather than looking at averages, we will consider individual treatment dynamics and study the heterogeneity among patients [6], using data from the German cohort [2]. As described above, in many patients, the dynamics appear to begin with a relatively fast phase of exponential decline. In some patients, this decline continues for the duration of the study. In most patients, however, this fast phase of decline is followed by a slower phase of exponential decline. Finally, in many patients, BCR-ABL transcript numbers resurge subsequently. Therefore, in most patients, there are three important slopes if the dynamics of BCR-ABL numbers are documented on a log-scale: (1) The slope of the fast decline phase, (2) the slope of the slower decline phase, and (3) the slope of the eventual rise. The value of these slopes was quantified by fitting a three phase exponential growth/decline model to the data, using non-linear least squares regression (Fig. 3.5a). A significant positive correlation was found between the slopes of the fast and the slow phases of decline. A significant negative correlation was found between the slopes of the slow phase of decline and the re-growth. Finally, no significant correlation was found between the slope of the fast phase of decline and the re-growth.

The negative correlation between the slopes of the slow phase of decline and the rise implies that both must probably be influenced by the presence of drug-resistant mutants [7–13]. The rise of drug-resistant mutants can slow down the decline of BCR-ABL transcript numbers, and is responsible for the eventual rise of BCR-ABL transcripts. The fact that the slopes of the slow and fast decline phases are positively correlated means that these two phases share a common mechanism of cell death that varies over time in strength. One such factor could be the immune system, which has been suggested to play a role in CML therapy before [14–16]. In particular, T cell responses (both CD4+ and CD8+) have been implicated.

3.5 Role of Immunity and Resistance in Driving Treatment Dynamics

The immune system is a complex entity that consists of many different components. Its main purpose of existence is to protect the organism from pathogens. However, the immune system is also thought to play a role in fighting tumor cells, thus contributing to protection against carcinogenesis. The immune system can roughly be subdivided into innate and adaptive immunity. Innate immunity involves non-specific mechanisms by which the organism fights intruders, ranging from barriers to infection (such as the skin), to fever that provides a suboptimal environment for pathogens, to cells that are involved in the removal of pathogens. The adaptive immune response, on the other hand, includes cells that specifically recognize pathogens and remove them. Therefore, this branch of the immune system is also called specific immunity. It plays a crucial part in clearing and controlling infections, which typically cannot be achieved by innate immunity alone. The main effectors of the adaptive immune response are antibodies and killer T cells. Killer T cells are thought to be particularly important for removing tumor cells. They typically kill cells that display foreign proteins, which can be either pathogen-infected cells or tumor cells displaying aberrant, mutated proteins. They are also called cytotoxic T lymphocytes (CTL), or CD8+ T cells. Both antibody and CTL response are activated by so called helper T cells or CD4+ T cells, which typically do not have direct effector function themselves. Specific immune cells exist at very low numbers before they have seen the protein they are specific to (also called the antigen). Upon encounter with the antigen, the

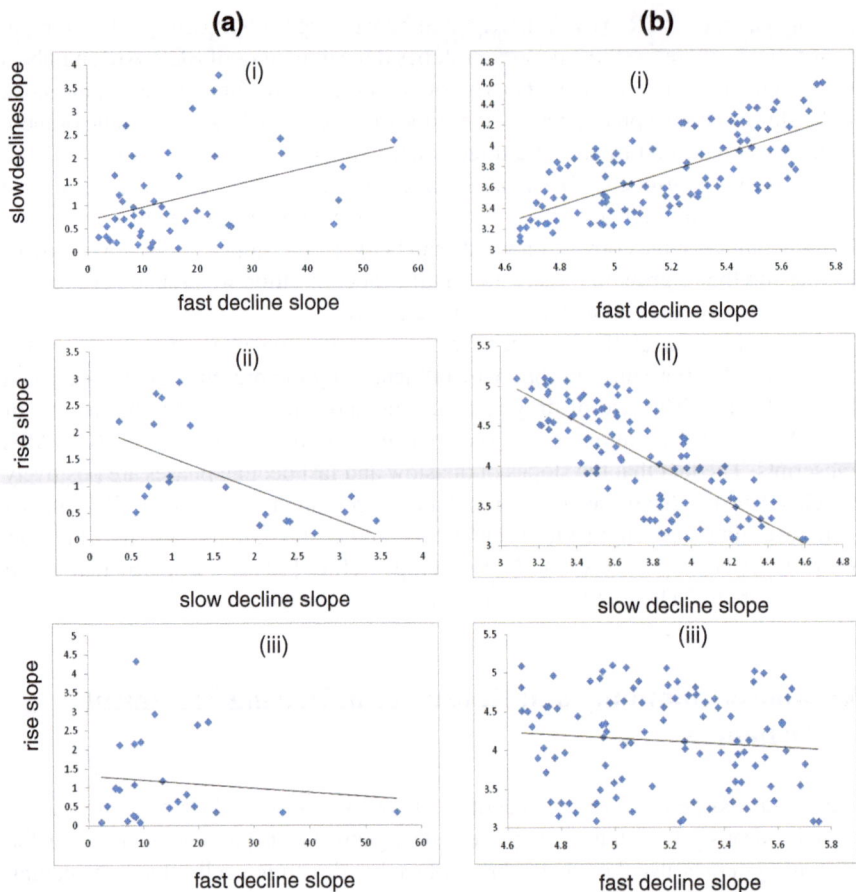

Fig. 3.5 Correlations between the slopes of the fast decline, slow decline, and the re-growth of BCR-ABL transcript numbers. (**a**) Correlations found in the clinical data from the German cohort of the IRIS study [2]. For (i) and (ii), the correlation is significant ($p = 0.009$ and $p = 0.003$, respectively), but not for (iii), $p = 0.59$. (**b**). Corresponding correlations as predicted by computer simulations. Simulations were run 100 times, varying randomly both the rate of specific immune proliferation, c, and the growth rate of the resistant cancer cells, r. For (i) and (ii), the correlations are significant ($p < 0.0001$ in each case), while the correlation is not significant for (iii), $p = 0.32$. The remaining parameters of the model (chosen for illustrative purposes) were: $\delta = 1$; $p = 3$; $b = 0.5$; $q = 0.01$; $\epsilon = 1$; $\eta = 1$; (units are year^{-1}). $x_0 = 100$; $y_0 = 0.001$; $z_0 = 0.5$

specific immune cell population expands through cell division and initiates effector function.

Data indicate that the level of CTL responses against CML is low before treatment, rises as treatment is administered, and declines again as the BCR-ABL transcript numbers fall to low levels [14–16]. These are similar immune response dynamics as observed in human immunodeficiency virus (HIV) and hepatitis C virus (HCV) infected patients that receive anti-viral drug therapy [17–19], and can be explained in the same way. While the cancer cells impair immunity, treatment reduces the amount of impairment, leading to a rise of immunity. As the number of cancer cells declines, withdrawl of antigenic stimulation causes a drop in immune responses. Such impairment dynamics have been explored extensively with mathematical models in the context of viral infections [20–22], and subsequently also in the context of CML [15]. The following will use a basic mathematical model that captures these assumptions. Model fitting to individual patient data demonstrates consistency with clinical observations, and the model further predicts the correlations found in the data.

A mathematical model is adapted that was previously published in the context of viral infections [20] and that captures the necessary assumptions. The model contains two variables: a growing virus population that has the ability to impair the immune system, and a specific immune response. In the present context, we are interested in CML and thus consider a population of growing CML cells that have the potential to impair immunity, and an immune response that reacts against CML antigens. The equations are the same and are given by:

$$dx/dt = Lx(1 - x/k) - dx - pxz, \qquad (3.3)$$
$$dz/dt = cxz/[(z + \eta)(x + \varepsilon)] - qxz - bz. \qquad (3.4)$$

The CML population is denoted by x, while the immune cell population is denoted by z. The CML population grows logistically. That is, at low numbers of cells, growth is exponential while growth slows down as the number of cells increases. The tumor cells die with a rate d and become removed by the immune response with a rate p. The immune response expands upon antigenic stimulation by tumor cells with a rate c. Immune expansion saturates if the number of tumor cells and the number of immune cells are high. The tumor cells impair immunity with a rate q. Finally, immune cells decay in the absence of antigenic stimulation with a rate b. The mathematical model is part of a general class of models that share common properties [20]. Results obtained from the model are therefore not dependent on the particular mathematical formulation used, but are robust. The equations that describe the immune response make the very basic assumptions that cancer cells can both stimulate the specific immune cells to proliferate, and impair the response. Because of the general nature of the model, this can be applied to any branch of the adaptive immune system including the T cell responses that are thought to play a role during CML therapy [14–16]. The basic model is characterized by the following outcomes, assuming that the degree of stimulation of immune cells is strong enough to drive immune expansion: (i) An immune response is successfully established and the system converges to a stable equilibrium in which the cancer is controlled at relatively low levels. (ii) The immune response goes extinct, and the cancer cell population converges to an equilibrium characterized by a large number of tumor cells. If the degree of immune impairment is relatively low, only the cancer control outcome is stable. If the degree of immune impairment lies above a threshold, both outcomes are stable. To which outcome the system converges depends in the initial conditions, with a high initial number of immune cells and a low initial number of tumor cells promoting the cancer control outcome.

For the current purposes, this general model was modified to include two sub-populations of cancer cells: drug sensitive and drug resistant cells. Denoting drug sensitive tumor cells by x and drug resistant tumor cells by y, the model is given by the following ordinary differentialequations.

$$dx/dt = -\delta x - pxz, \tag{3.5}$$

$$dy/dt = ry - pyz, \tag{3.6}$$

$$dz/dt = c(x + y)z/[(z + \eta)(x + y + \varepsilon)] - q(x + y)z - bz, \tag{3.7}$$

where we denoted

$$L - D - H \equiv -\delta < 0, \quad L - D \equiv r > 0.$$

Because drug-sensitive cells, x, are susceptible to therapy, this population of cells is assumed to decline during treatment with a rate δ. As before, immune responses contribute to cell death with a rate p. On the other hand, drug-resistant cells, y, are not susceptible to therapy and are thus assumed to expand exponentially during therapy with a rate r. Because the initial growth phase of the resistant CML cells is interesting in the current context, we ignore growth saturation at high numbers of cells for simplicity. As with drug-sensitive cells, drug-resistant cells are removed by immune responses with a rate p. The equation for immune responses is the same as in the simple model described previously. However, both antigenic stimulation and immune impairment are now proportional to the number of drug-sensitive and drug-resistant cells, $x + y$.

The model can give rise to treatment dynamics that can describe diverse clinical data well, shown by nonlinear least squares fits of the model to selected patient data (Fig. 3.6). Figure 3.6a shows a response that involves a faster, followed by a slower phase of decline, eventually leading to a resurgence of the cancer. Immunity rises as the number of CML cells starts to decline during treatment because of reduced immune impairment. This accounts for the fast phase of CML decline. Immunity subsequently falls due to lack of antigenic stimulation. This, together with the rise of a resistant mutant, accounts for the slower phase of decline. The rise of the resistant mutant eventually leads to resurgence of the cancer. Figure 3.6b shows the same kind of profile without eventual resurgence of the cancer. In this case, a resistant mutant is not present. Figure 3.6c shows a single phase exponential decline of CML cells, while Fig. 3.6d shows an exponential decline followed by re-growth of the cancer. In these cases, immunity does not expand during the treatment dynamics according to the model, correlating with a significantly slower decline rate of the cancer. In Fig. 3.6c, no resistant mutant exists, while in Fig. 3.6d, a resistant mutant grows during therapy.

To examine predicted correlations between the slopes, the simulation was run many times, randomly varying the growth rate of drug-resistant CML cells and the

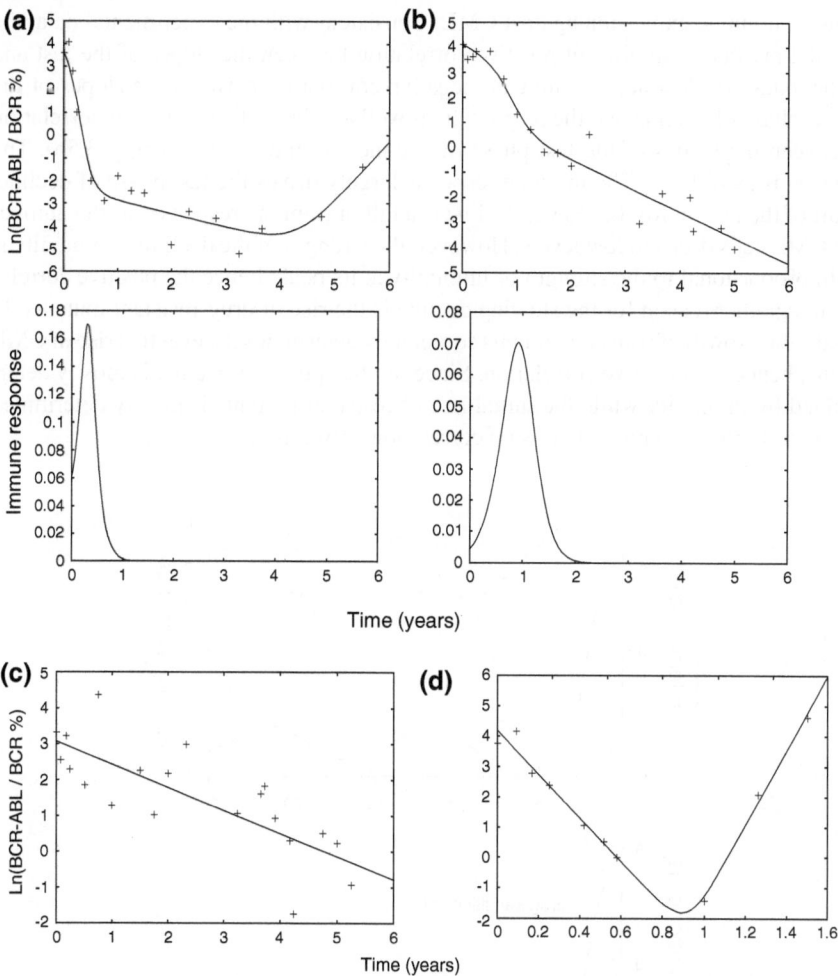

Fig. 3.6 Examples of individual treatment response data from the German cohort of the IRIS study, taken from [2]. The dots are the actual clinical data. The line is the model fit to the data, obtained by non-linear least squares regression. In (**a**) and (**b**) immune responses rise temporarily during therapy, contributing to the overall treatment dynamics. In (**c**) and (**d**), the model predicts that immune responses did not rise during therapy. Drug-resistant mutants play a role in (**a**) and (**d**), but not in (**b**) and (**c**). Model parameter values are as follows. (**a**) $r = 2.69$; $c = 17.04$; $\delta = 0.58$; $p = 80.3$; $b = 8.84$; $q = 0.10$; $\epsilon = 1$; $\eta = 1$; (units are year^{-1}). $x_0 = 35.53$; $y_0 = 2.73E - 5$; $z_0 = 0.06$. (**b**) $r = 0$; $c = 14.00$; $\delta = 1.04$; $p = 43.60$; $b = 8.97$; $q = 4.88E - 4$; $\epsilon = 2.85$; $\eta = 1$; (units are year^{-1}). $x_0 = 60.76$; $y_0 = 0$; $z_0 = 0.0044$. (**c**) $r = 0$; $c = 0$; $\delta = 0.65$; $p = 0$; $b = 0$; $q = 0$; $\epsilon = 0$; $\eta = 0$; (units are year^{-1}). $x_0 = 21.17$; $y_0 = 0$; $z_0 = 0$. (**d**) $r = 12.4$; $c = 0$; $\delta = 7.23$; $p = 0$; $b = 0$; $q = 0$; $\epsilon = 0$; $\eta = 0$; (units are year^{-1}). $x_0 = 65.92$; $y_0 = 9.61E - 7$; $z_0 = 0$

rate of immune expansion against CML. Consistent with the experimental data, the model predicts a significant positive correlation between the slopes of the fast and slow phase of decline, a significant negative correlation between the slopes of the slow phase of decline and the re-growth, as well as a lack of a significant correlation between the slopes of the fast phase of decline and the re-growth (Fig. 3.5b). The reason is as follows. The immune response largely drives the fast phase of decline. Part of the reason for the slower decline is a fall of immune responses as the number of CML cells drops to low levels. However, the strength of the declining immunity is still proportional to the strength of immunity at its peak, hence the positive correlation. Another reason for the slowing decline is the rise of drug-resistant mutants. In addition, growth of resistant mutants completely determines the eventual rise of CML cells, hence the negative correlation. Since the fast phase of decline is mostly determined by immunity while the initial rise of resistant mutants is mostly determined by their replication rate, there is no correlation between these slopes.

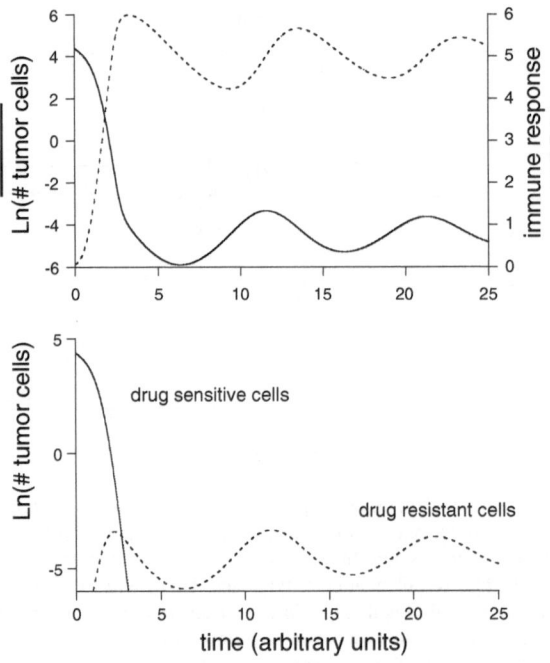

Fig. 3.7 Computer simulation showing the possibility that during therapy, the immune response is maintained at higher levels rather than dropping to insignificant levels, leading to long-term control of CML. The *lower panel* shows the CML dynamics separately for the populations of drug sensitive and drug-resistant cells. Parameters were chosen as follows: $r = 5$; $c = 5$; $\delta = 0.5$; $p = 1$; $b = 0.1$; $q = 0.01$; $\epsilon = 0.1$; $\eta = 1$; $x_0 = 80$; $y_0 = 2.7 \times 10^{-5}$; $z_0 = 0.06$. The units of the axes are arbitrary, as the parameter set was chosen for illustrative purposes and is not based on measured parameters that are currently unknown

3.6 Possible Role of Immune Stimulation for Long-Term Remission

If immune responses that arise during treatment are sustained, they could in principle suppress the population of drug-resistant cancer cells. In some patients, discontinuation of therapy after 2 years did not result in the relapse of the cancer [5]. It can be hypothesized that in these cases, immunity became fully established during therapy rather than rising only temporarily. This possibility is demonstrated by computer simulation in Fig. 3.7. Establishment of sustained immune responses during therapy of immunosuppressive diseases has been found to occur in some experimental HCV and SIV therapy regimes [17, 23].

3.7 Summary

This chapter discussed the basic CML dynamics during targeted treatment with imatinib, focusing on the decline of the BCR-ABL transcript levels. The typical patterns of the decline dynamics were discussed, and different explanations were reviewed, based on mathematical modeling. Emphasis was placed on the distinction between cancer stem cells and more differentiated cells, and on the role of the activation status of CML stem cells for susceptibility to treatment. The dynamics of CML stem cell quiescence and activation were shown to be able to account for a variety of treatment response patterns that have been documented. In addition, heterogeneity in the decline dynamics of CML cells among different patients was discussed, and could be accounted for by different dynamics of CML specific immune responses and the occurrence of drug resistant cell clones. This provides a transition to the following chapters, which discuss the evolutionary dynamics of drug-resistant mutants in detail, the biggest obstacle to long-term control of the disease.

Problems

3.1 Study the system of ODEs (3.1–3.2) by writing down the general solution, and derive the expressions for $\lambda_{\pm}$.

3.2 Find fixed points of system (3.5 – 3.7). Perform a linear stability analysis of the trivial fixed point, $(0, 0, 0)$.

3.3 Numerical project
Investigate the behavior of system (3.5 and 3.7) numerically. Use any software or a programing language to find numerical solutions of the system under different initial conditions and under different parameter values.

References

1. Michor, F., Hughes, T.P., Iwasa, Y., Branford, S., Shah, N.P., Sawyers, C.L., Nowak, M.A.: Dynamics of chronic myeloid leukaemia. Nature **435**(7046), 1267–1270 (2005)
2. Roeder, I., Horn, M., Glauche, I., Hochhaus, A., Mueller, M.C., Loeffler, M.: Dynamic modeling of imatinib-treated chronic myeloid leukemia: functional insights and clinical implications. Nat. Med. **12**(10), 1181–1184 (2006)
3. Reya, T., Morrison, S.J., Clarke, M.F., Weissman, I.L.: Stem cells, cancer, and cancer stem cells. Nature **414**(6859), 105–111 (2001)
4. Teschl, G.: Ordinary Differential Equations and Dynamical Systems, vol. 140. American Mathematical Society, Providence (2012)
5. Rousselot, P., Huguet, F., Rea, D., Legros, L., Cayuela, J.M., Maarek, O., Blanchet, O., Marit, G., Gluckman, E., Reiffers, J., Gardembas, M., Mahon, F.X.: Imatinib mesylate discontinuation in patients with chronic myelogenous leukemia in complete molecular remission for more than two years. Blood **109**(1), 58–60 (2007)
6. Wodarz, D.: Heterogeneity in chronic myeloid leukaemia dynamics during imatinib treatment: role of immune responses. Proc. Biol. Sci. **277**(1689), 1875–1880 (2010)
7. Deininger, M.W., Druker, B.J.: Specific targeted therapy of chronic myelogenous leukemia with imatinib. Pharmacol. Rev. **55**(3), 401–423 (2003)
8. Druker, B.J.: Overcoming resistance to imatinib by combining targeted agents. Mol. Cancer. Ther. **2**(3), 225–226 (2003)
9. Druker, B.J.: Imatinib as a paradigm of targeted therapies. Adv. Cancer. Res. **91**, 1–30 (2004)
10. Gambacorti-Passerini, C.B., Gunby, R.H., Piazza, R., Galietta, A., Rostagno, R., Scapozza, L.: Molecular mechanisms of resistance to imatinib in philadelphia-chromosome-positive leukaemias. Lancet Oncol. **4**(2), 75–85 (2003)
11. Gorre, M.E., Mohammed, M., Ellwood, K., Hsu, N., Paquette, R., Rao, P.N., Sawyers, C.L.: Clinical resistance to sti-571 cancer therapy caused by bcr-abl gene mutation or amplification. Science **293**(5531), 876–880 (2001)
12. Nardi, V., Azam, M., Daley, G.Q.: Mechanisms and implications of imatinib resistance mutations in bcr-abl. Curr. Opin. Hematol. **11**(1), 35–43 (2004)
13. Shah, N.P., Tran, C., Lee, F.Y., Chen, P., Norris, D., Sawyers, C.L.: Overriding imatinib resistance with a novel abl kinase inhibitor. Science **305**(5682), 399–401 (2004)
14. Chen, C.I., Maecker, H.T., Lee, P.P.: Development and dynamics of robust t-cell responses to cml under imatinib treatment. Blood **111**(11), 5342–5349 (2008)
15. Kim, P.S., Lee, P.P., Levy, D.: Dynamics and potential impact of the immune response to chronic myelogenous leukemia. PLoS Comput Biol **4**(6), e1000,095 (2008)
16. Wang, H., Cheng, F., Cuenca, A., Horna, P., Zheng, Z., Bhalla, K., Sotomayor, E.M.: Imatinib mesylate (sti-571) enhances antigen-presenting cell function and overcomes tumor-induced cd4+ t-cell tolerance. Blood **105**(3), 1135–1143 (2005)
17. Barnes, E., Harcourt, G., Brown, D., Lucas, M., Lechner, F., Phillips, R., Dusheiko, G., Klenerman, P.: In: Effects of combination therapy for hepatits c virus on virus-specific immune responses. (2002, Hepatology in press)
18. Kalams, S.A., Buchbinder, S.P., Rosenberg, E.S., Billingsley, J.M., Colbert, D.S., Jones, N.G., Shea, A.K., Trocha, A.K., Walker, B.D.: Association between virus-specific cytotoxic t-lymphocyte and helper responses in human immunodeficiency virus type 1 infection. J. Virol. **73**(8), 6715–6720 (1999)
19. Ogg, G.S., Jin, X., Bonhoeffer, S., Moss, P., Nowak, M.A., Monard, S., Segal, J.P., Cao, Y., Rowland-Jones, S.L., Hurley, A., Markowitz, M., Ho, D.D., McMichael, A.J., Nixon, D.F.: Decay kinetics of human immunodeficiency virus-specific effector cytotoxic t lymphocytes after combination antiretroviral therapy. J. Virol. **73**(1), 797–800 (1999)
20. Komarova, N.L., Barnes, E., Klenerman, P., Wodarz, D.: Boosting immunity by antiviral drug therapy: a simple relationship among timing, efficacy, and success. Proc. Natl. Acad. Sci. U S A **100**(4), 1855–1860 (2003)

21. Wodarz, D.: Helper-dependent vs. helper-independent ctl responses in hiv infection: implications for drug therapy and resistance. J. Theor. Biol. **213**(3), 447–459 (2001)
22. Wodarz, D., Nowak, M.: Specific therapy regimes could lead to long-term control of hiv. Proc. Natl. Acad. Sci. USA **96**, 14,464–14,469 (1999)
23. Lifson, J., Rossio, J., Arnaout, R., Li, L., Parks, T., Schneider, D., Kiser, R., Coalter, V., Walsh, G., Imming, R., Fischer, B., Flynn, B., Nowak, M., Wodarz, D.: Containment of siv infection: cellular immune responses and protection from rechallenge following transient post-inoculation antiretroviral treatment. J. Virol. **74**, 2584–2593 (2000)

Chapter 4
Stochastic Modeling of Cellular Growth, Treatment, and Resistance Generation

Abstract In this chapter we develop some useful mathematical tools that allow us to model the growth of cancer, the generation of resistance, and the effect of multidrug treatment. The methodology is stochastic, based on the description of a cellular colony as a birth–death process with mutations. This chapter provides mathematical foundation for all the subsequent chapters on small molecule inhibitor treatments and resistance generation. The less mathematically inclined readers may skip this chapter.

Keywords Stochastic process · Continuous time Markov process · Linear birth-death process · Divisions · Cell death · Mutations · Transformations · Mutation networks · Susceptible cells · Resistant cells · Partially resistant cells · Combination therapies · Drug-induced cell death · Cancer turnover · Kolmogorov forward equation · Probability generating function · PDEs · Method of characteristics · Probability of treatment success · Treatment stages

4.1 Introduction

The first stochastic model of drug resistance was created by Goldie and Coldman [1], who developed a whole new approach to mathematical treatment of resistance in their subsequent work, see e.g., [2–7]. A number of important theoretical and numerical results have been obtained by the authors for the case of one and more drugs. Since this ground-breaking work, a lot of mathematical models of drug resistance in cancer have been proposed. Several models, including stochastic branching models for stable and unstable gene amplification and its relevance to drug resistance, were explored by [8–12]. Methods of optimal control theory were used to analyze drug dosing and treatment strategies [13–18] (for a review of the optimal control theory in chemotherapy see [19]). Models for tumor growth incorporating age-structured cell cycle dynamics, in application to chemotherapy scheduling, have been developed

N. L. Komarova and D. Wodarz, *Targeted Cancer Treatment in Silico*,
Modeling and Simulation in Science, Engineering and Technology,
DOI: 10.1007/978-1-4614-8301-4_4, © Springer Science+Business Media New York 2014

by [20, 21]. Mechanistic mathematical models developed to improve the design of chemotherapy regimes are reviewed in [22]. Jackson and Byrne [23] extended an earlier PDE model of [24] to study the role of drug resistance and vasculature in tumors' response to chemotherapy; in this class of spatial models, the tumor is treated as a continuum of different types of cells, which include susceptible and resistant cells. Another class of models is based on the Luria-Delbruck mutation analysis [25–27].

In this and subsequent chapters we formulate a set of both stochastic and deterministic models for multi-drug resistance and investigate the dependence of treatment outcomes on the initial tumor load, mutation rates, the turnover rate of cancerous cells, and the treatment strategy. The main goal is to elucidate the general principles of the emergence and evolution of resistant cells inside the tumor, before and after the start of treatment. The stochastic model follows the tradition of [2, 7, 28], and takes this classical work a step further. In particular, it has been possible to include a nonzero death rate for cancer cells and still obtain analytical results. Other extensions include the studies of cross-resistance, quiescence/cycling transitions, and a variety of different treatment strategies.

4.2 The Basic Model of Cancer Growth and Generation of Mutations

4.2.1 The Concept: A Birth–Death Process with Mutations

The aim is to describe the dynamics of birth, death, and mutations in a colony of cells [29, 30]. We assume that there are four different processes that can take place in the colony, see Fig. 4.1. A death will lead to the number of cells of the given type decreasing by one. Faithful reproduction will increase the number of cells of the given type. Reproduction with a mutation will result in an increment in the number of cells of the mutant type. Transformation will decrease the number of cells of a given type and simultaneously increase the number of cells of the mutant type.

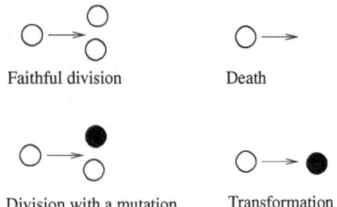

Fig. 4.1 The four basic processes: death, faithful and unfaithful division, and transformation. For simplicity, only two cell types are included: wild type and transformed cells. Wild type cells are depicted as empty circles, and transformed cells as filled circles

Such transformations can be caused by environmental factors or can be a direct consequence of treatment by mutagenic drugs.

Because of mutations, there may be cells of different types in a colony. Each cellular type (phenotype) in this model is characterized by its resistance properties. For instance, if we consider one drug, then a cell can be either resistant or susceptible to this drug. The system has only two types, and the state space consists of vectors (i, j), where $i, j \geq 0$ are integer numbers denoting the number of susceptible and resistant cells respectively. We assume that resistance to the drug is acquired by means of a mutation. This process is reflected by the following (very simple) network, $x_0 \to^u x_1$. Here u is the mutation rate, that is, the probability to create a resistant mutant upon division, and the index k in notation x_k characterizes susceptibility, with $k = 0$ meaning susceptible, and $k = 1$ not susceptible (resistant) to the drug.

We can generalize this process of resistance generation to the case of m different drugs [29, 30]. Each cell can acquire resistance to each of the drugs, by means of a certain mutation. There are m types of mutations, with rates $u_1, \ldots, u_m$. Each mutation event corresponding to rate u_i leads to a phenotype resistant to drug i. We assume that resistance to one drug does not imply resistance to another drug (in general, this is not always true and cross-resistance is often observed, which is addressed later in this book). In order to develop resistance to all m drugs, a cell must accumulate m mutations. Simple combinatorics suggests that there can be up to

$$n = 2^m - 1$$

different resistant phenotypes. The total number of types is 2^m, and the system's state is characterized by an integer-valued vector of length 2^m. In particular, there are $\binom{m}{k}$ phenotypes resistant to $k \leq m$ drugs. Figure 4.2 illustrates the mutation network for $m = 3$. We label each phenotype by a binary number of length m, where "1" indicates resistance to the drug corresponding to its position and "0" indicates susceptibility. For example, if $m = 3$, type 101 is resistant to drugs number 1 and 3 and it is susceptible to drug number 2.

We will model the dynamics by using a continuous time, discrete state space Markov birth–death process on a combinatorial mutation network. For an infinitesimal time increment, Δt, the probability for a cell to undergo each of the four basic processes (death, faithful division, mutation, and transformation) is proportional to

Fig. 4.2 Mutation diagram corresponding to three drugs. The binary number, s, for each node describes the resistance properties of the corresponding phenotype: 0 stands for "susceptible" and 1 for "resistant"

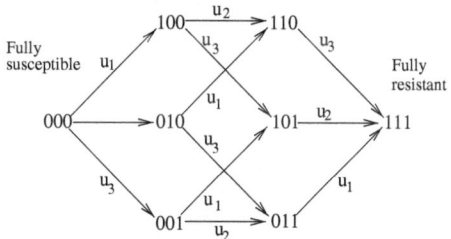

Δt; the probability of two events occurring simultaneously (within Δt) is neglected. Different events will occur with different probabilities (or rates), depending on whether treatment is applied. We denote by L_s the birth rate of type s, and by d_s its death rate.

In general, the death rate of cells is comprised of their *natural death date*, D_s, which is equal to the death rate in the absence of treatment, and possibly a *drug-induced death rate*, H_s. We have

$$d_s = D_s + H_s.$$

For the wild-type cancer cells we assume that the corresponding rates satisfy $L_0 > D_0$, so that the colony grows in the absence of treatment. The ratio $0 \leq D_0/L_0 < 1$ defines the *turnover* of cancer cells. The values $D_0/L_0 \ll 1$ correspond to low turnover, low death cancers, whereas values $D_0/L_0 \approx 1$ describe extremely high-turnover, slow-growth cancers.

In addition to the natural death rate, cells can be subject to an additional, drug-induced death rate, H_s. There are many ways to model drug-induced death rates. For instance, we could assume that each drug a cell is susceptible to increases its death rate by a certain amount, so the more drugs a cell is resistant to, the higher its death rate. To model partial resistance, we could assume that even "maximally resistant" cells have a residual drug-induced death rate. However, here we will adopt a simpler framework. We assume that if a cell is resistant to *all* the drugs applied, then its drug-induced death rate is simply zero. On the other hand, if a cell is susceptible to at least one of the drugs, then its drug-induced death rate is equal to a constant, H, which has the meaning of the intensity of therapy. The more complicated scenarios described above can also be incorporated in the model, see e.g., [31].

Here is a formal description of all the processes [29, 30]. Each phenotype "A^s" with some $0 \leq s \leq 2^m$, is represented as a node of a mutation network with arrows coming in and out, see figure 4.3. We assume that in time interval Δt, the following events can occur with each phenotype "A^s":

Fig. 4.3 An example of a vortex in a mutation diagram

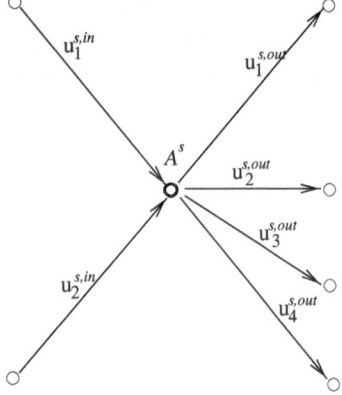

- With probability $L_s(1 - \sum_j u_j^{s,\text{out}})\Delta t$ a cell of type "A^s" reproduces, creating an identical copy of itself.
- For each outgoing arrow, with probability $L_s u_j^{s,\text{out}}$ a cell of type "A^s" reproduces with a mutation, creating a cell of type "$A_j^{s,\text{next}}$", for all j.
- With probability $d_s \Delta t$ a cell of type "A^s" dies.
- For each outgoing arrow, with probability $K_s \alpha_j^{s,\text{out}}$ a cell of type "A^s" is transformed into a cell of type "$A_j^{s\text{next}}$", for all j.

In this model, we have $d_s = D_s + H_s$. Coefficients H_s are individual drug-induced death rates, and K_s are coefficients responsible for the transformation rates; both in some sense reflect the strength (intensity) of therapy, whereas K_s also reflects how mutagenic the drugs are. In all the applications described here we only consider non-mutagenic drugs, such that $K_s = 0$.

We start with a given number of cells of type "A^0", and follow the process until the first cell of type "A^n" has been created. We would like to calculate the probability, $P_n(t)$, that a cell of type "A^n" has been created as a function of time.

4.2.2 Stochastic Description: The Example of One Drug

Before we introduce the general methodology, let us consider a specific example of a one-drug treatment. The states of the system are characterized by integer-values vectors, (i_0, i_1), where i_0 is the number of cells susceptible to the drug and i_1 is the number of cells resistant to the drug. Let us denote by $\varphi_{i_0,i_1}(t)$ the probability to be in state (i_0, i_1) at time t. Let us consider the processes of cell divisions, death, and mutations from type 0 to type 1 with probability u per cell division. This function φ_{i_0,i_1} satisfies the following Kolmogorov forward equation:

$$
\begin{aligned}
\varphi_{i_0,i_1} = {}& L_0(1-u)\varphi_{i_0-1,i_1}(i_0-1) + L_1\varphi_{i_0,i_1-1}(i_1-1) \\
& + L_0 u \varphi_{i_0,i_1-1} i_0 \\
& + d_0 \varphi_{i_0+1,i_1}(i_0+1) + d_1 \varphi_{i_0,i_1+1}(i_1+1) \\
& + -\varphi_{i_0,j_0}(L_0 + L_1 + d_0 + d_1),
\end{aligned} \tag{4.1}
$$

where the terms on the first line represent divisions of both types of cells, the second line corresponds to mutations creating resistant cells, the third line is death of both types of cells, and the last line corresponds to the possibility of no change (the negative term). To proceed, it is convenient to define the probability generating function,

$$
\Psi(\xi_0, \xi_1; t) = \sum_{i_0,i_1} \varphi_{i_0,i_1}(t)\xi_0^{i_0}\xi_1^{i_1}.
$$

In order to derive an equation satisfied by this function, we multiply Eq. (4.1) by $\xi_0^{i_0}\xi_1^{i_1}$ and sum over i_0 and i_1. Then the left hand side becomes $\frac{\partial \Psi}{\partial t}$. On the right hand side we have several types of terms. For example, the first term can be written as

$$\sum_{i_0,i_1} \varphi_{i_0-1,i_1}(i_0-1)\xi_0^{i_0}\xi_1^{i_1} = \sum_{k,i_1} \varphi_{k,i_1}k\xi_0^{k+1}\xi_1^{i_1},$$

where $k = i_0 - 1$. Next we notice this can be rewritten as

$$\xi_0^2 \sum_{k,i_1} \varphi_{k,i_1}k\xi_0^{k-1}\xi_1^{i_1} = \xi_0^2 \frac{\partial}{\partial \xi_0} \sum_{k,i_1} \varphi_{k,i_1}\xi_0^k\xi_1^{i_1} = \xi_0^2 \frac{\partial}{\partial \xi_0}\Psi.$$

Similarly, the other terms can be rewritten in terms of partial derivatives of the function Ψ with respect to ξ_0 or ξ_1. We obtain the following equation:

$$\frac{\partial \Psi}{\partial t} = \frac{\partial \Psi}{\partial \xi_0}(L_0(1-u)\xi_0^2 + d_0 - \xi_0(L_0 + d_0 - L_0 u\xi_1)) + \frac{\partial \Psi}{\partial \xi_1}(L_1\xi_1^2 + d_1 - \xi_1(L_1+d_1)).$$

$$(4.2)$$

In the next section, we will generalize this methodology to multiple drugs and then show how it can be used to answer important questions related to drug treatments of cancer. Before we go on, we will explain the general usefulness of the approach outlined here.

The probability function, $\varphi_{i_0,i_1}(t)$ contains all the "microscopic" information about the stochastic process of interest. In fact, it contains "too much" information, and often one could make shortcuts to extract only the "useful" information needed to answer a specific question. One such shortcut is demonstrated by the method outlined above and generalized below. The calculations leading to Eq. (4.2) demonstrate the convenience of the probability generating function description: instead on an infinite system of coupled Eq. (4.1) for the original probability function, $\varphi_{i_0,i_1}(t)$, we obtained a single first order PDE for the probability generating function. This PDE can be solved by the method of characteristics (as reviewed below). The solutions can be related to important characteristics of the process such as the probability of treatment success/failure.

4.2.3 The Generating Function Description

Next, we present a general approach to describe the stochastic evolution of a cellular colony with mutations. For a comprehensive introduction to stochastic processes, we refer the reader to [32]. In the general case of m drugs, let us introduce the function

$$\varphi_{i_0,\ldots,i_n}(t),$$

the probability to have i_s cells of type A^s at time t, where $0 \le s \le n = 2^m$ are binary numbers. We can write down the Kolmogorov forward equation,

$$\dot{\varphi}_{i_0,\ldots,i_n} = \sum_{s=0}^{n} Q\{A^s\},$$ (4.3)

where $Q\{A^s\}$ is the contribution obtained from considering probabilities of reproduction and death of cell-type "A^s",

$$Q\{A^s\} = \varphi_{\ldots,i_s-1,\ldots}(i_s - 1)L_s\left(1 - \sum_j u_j^{s,\text{out}}\right) + i_s L_s \sum_j \varphi_{\ldots,i_s,\ldots,i_j-1,\ldots} u_j^{s,\text{out}}$$

$$+ \varphi_{\ldots,i_s+1,\ldots}(i_s + 1)d_s + K_s \sum_j \alpha_j^{s,\text{out}} \varphi_{\ldots,i_s+1,\ldots,i_j-1,\ldots}(i_s + 1)$$

$$- \varphi_{\ldots}i_s\left(L_s + d_s + K_s \sum_j \alpha_j^{s,\text{out}}\right).$$ (4.4)

We used the following short-hand notations: $\varphi_{\ldots}$ stands for $\varphi_{i_0,\ldots,i_n}$, and the only explicit subscripts indicate the indices which are different from $(i_0, \ldots, i_n)$. In Eq. (4.4), the first term is faithful reproduction, the second term represents all possible mutations, the third term is death, the forth term is transformation of cells, and the last term comes from the probability of no change.

It is convenient to define the probability generating function, $\Psi(\xi_0, \ldots, \xi_n; t)$,

$$\Psi(\xi_0, \ldots, \xi_n; t) = \sum_{i_0,\ldots,i_n} \varphi_{i_0,\ldots,i_n} \prod_{s=0}^{n} \xi_s^{i_s}.$$ (4.5)

This expression can be viewed as a transformation from discrete variables $i_0, \ldots i_n$ to continuous variables $\xi_0, \ldots, \xi_n$.

Let us multiply Eq. (4.3) by $\prod_{s=0}^{n} \xi_s^{i_s}$ and sum over all indices to obtain the equation for the generating function. The rule for rewriting various types of terms in terms of the generating function can be summarized as follows:

- Terms with $\varphi_{\ldots}i_s$ give $\xi_s \frac{\partial}{\partial \xi_s}$;
- Terms that multiply $\varphi_{\ldots i_s-1\ldots}(i_s - 1)$ give $\xi_s^2 \frac{\partial}{\partial \xi_s}$;
- Terms that multiply $\varphi_{\ldots i_s+1\ldots}(i_s + 1)$ give $\frac{\partial}{\partial \xi_s}$;
- Terms like $\varphi_{\ldots i_s \ldots i_j-1\ldots}i_s$ give $\xi_s \xi_j \frac{\partial}{\partial \xi_s}$;
- Terms with $\varphi_{\ldots i_s+1\ldots i_j-1}(i_s + 1)$ give $\xi_j \frac{\partial}{\partial \xi_s}$.

It is convenient to introduce the following shorthand notations,

$$u^{s,\text{out}} = \sum_j u_j^{s,\text{out}}, \quad \alpha^{s,\text{out}} = \sum_j \alpha_j^{s,\text{out}}.$$

Then the function $\Psi(\xi_0, \ldots, \xi_n; t)$ satisfies the following hyperbolic partial differential equation:

$$\frac{\partial \Psi}{\partial t} = \sum_s \frac{\partial \Psi}{\partial \xi_s} \left[\xi_s^2 L_s \left(1 - u^{s,\text{out}} \right) + d_s + \xi_s L_s \sum_j \xi_j u_j^{s,\text{out}} \right.$$

$$\left. + K_s \sum_j \xi_j \alpha_j^{s,\text{out}} - \left(L_s + d_s + K_s \alpha^{s,\text{out}} \right) \xi_s \right]. \quad (4.6)$$

The fact that this is a first order equation follows from the assumption of linearity of the underlying birth-death process. Nonlinear processes lead to higher order equations.

4.2.4 Treatment Stages and the Method of Characteristics

The first-order PDE (4.6) can be solved by the standard method of characteristics, see e.g., [33], Chap. 8. The equations for characteristics for Eq. (4.6) are given by:

$$\dot{\xi}_s = L_s \left(1 - u^{s,\text{out}} \right) \xi_s^2 + \left[L_s \sum_j u_j^{s,\text{out}} \xi_j - \left(L_s + d_s + K_s \alpha^{s,\text{out}} \right) \right] \xi_s$$

$$+ K_s \sum_j \xi_j \alpha_j^{s,\text{out}} + d_s), \quad 0 \le s \le n. \quad (4.7)$$

Let us assume that at $t = 0$, we have M_0 cells resistant to zero drugs, so that $\varphi_{M_0,0,\ldots,0}(0) = 1$. From definition (4.5) we have

$$\Psi(\xi_0, \ldots, \xi_n; 0) = \xi_n^{M_0}.$$

Suppose that we want to obtain expression for $\Psi(\bar{\xi}_0, \ldots, \bar{\xi}_n; \bar{t})$ by the method of characteristics, where $\bar{\xi}_0, \ldots, \bar{\xi}_n$ and $\bar{t}$ are some fixed values. For time $\bar{t}$, we have

$$\Psi(\bar{\xi}_0, \ldots, \bar{\xi}_n; \bar{t}) = \xi_n(\bar{t})^{M_0},$$

where the function $\xi_n(t)$ is a solution of system (4.7) which satisfies the initial conditions,

$$\xi_j(0) = \bar{\xi}_j, \quad 0 \le j \le n. \quad (4.8)$$

Let us denote by t_* the time when treatment begins. The timing of treatment, that is, the value t_*, is related to the tumor size at the start of treatment, N. In the simplest case we can assume a deterministic relationship,

$$N = M_0 e^{(L_0 - D_0)t_*}, \quad (4.9)$$

where all the types in the absence of treatment are assumed to divide and die with the same rates given by L_0 and D_0. This is an approximation, which is made throughout this book. Its consequences are explored in detail in [34].

In general, coefficients in system (4.7) are time-dependent, because the death rate of cells, d_s, depend on whether treatment is applied or not. In the simplest case, we have the following 2-stage process: for $0 < t < t_*$ no treatment is applied, such that $d_s = D_s$. We call this regime the pretreatment stage. At time $t = t_*$, therapy starts and continues for all values of $t \geq t_*$, such that $d_s = D_s + H_s$. This is the treatment stage. More complex processes consisting of more than two stages are considered in Chaps. 7 and 9. We would like to calculate the function $\Psi(\bar{\xi}_0, \ldots, \bar{\xi}_n; t_* + \bar{t})$, that is, the characteristic function at time $\bar{t}$ after the start of therapy. In this case we have two sets of equations for characteristics of type (4.7), one with "pretreatment coefficients" and the other with "treatment coefficients".

In order to solve this problem, we can employ the following simple but super-ficially counterintuitive algorithm. First, we solve the system of ODEs for charac-teristics with *treatment* coefficients, with the initial conditions given by $\xi_0(0) = \bar{\xi}_0, \ldots, \xi_n(0) = \bar{\xi}_n$, and obtain the solution $\xi_0(\bar{t}), \ldots, \xi_n(\bar{t})$. Next, we use these val-ues as the initial condition for the system with *pretreatment* coefficients, and get the solution of that at time t_*, $\xi_0(t_*), \ldots, \xi_n(t_*)$. The result is then given by

$$\Psi(\bar{\xi}_0, \ldots, \bar{\xi}_n; \bar{t}) = \xi_n(t_*)^{M_0}.$$

The reason for this time-treversal is the equations for characteristics use to trace the trajectories back to their initial condition. This is consistent with the standard 1st order PDE techniques.

4.2.5 Probability of Extinction and Treatment Success

The function $\Psi(0, 1, \ldots, 1; t)$ has the meaning of the probability that at time t, no cells of type "A^n" exist. We will call this quantity "the probability of non-production"(of resistance); it is given by

$$P_{\text{non-prod}}(t) = \Psi(0, 1, \ldots, 1; t) = \xi_0(t)^{M_0},$$

where M_0 is the initial number of wild-type cells, and functions $\xi_i(t)$ satisfy system (4.7) with initial conditions

$$\xi_0 = 0, \quad \xi_i = 1, \quad 1 \leq i \leq n.$$

Probability of extinction, that is, the probability that there are zero cells of each type at time t, can be defined as

$$P_{\text{ext}} = \Psi(0, 0, \ldots, 0; t) = (\xi_0(t))^{M_0},$$

where $\xi_n(t)$ is the solution of the system (4.7) with different initial conditions,

$$\xi_0(0) = \xi_1(0) = \ldots = \xi_n(0) = 0.$$

Note that in the regime of treatment, the probability of extinction, as $t \to \infty$, coincides with the long-time limit of the probability of non-production of a resistant mutant,

$$\lim_{t \to \infty} P_{\text{ext}} = \lim_{t \to \infty} P_{\text{non-prod}}.$$

The function $\Psi(0, \ldots, 0; t)$ is a monotonically increasing function, which equals zero for $t = 0$, and steadily approaches a horizontal asymptote. The function $\Psi(0, 1, \ldots, 1; t)$ is not necessarily monotonic. It starts at one for $t = 0$, can drop to a minimum and then climb up and approach the same asymptote from below. In other scenarios, it can be a monotonically decreasing function of time (e.g., for $D = 0$). The reason for a nonmonotonic behavior is this. As treatment starts at $t = 0$, there are no mutants, and then some resistant mutants may be produced quickly, while the population of cells is still large. Later on, the susceptible population decreases dramatically, so no new resistant mutants are produced, but there is a natural death rate for the mutants which may lead to accidental extinction of the mutant colony before it reaches a significant size. Once the resistant colony grows, the chance of its spontaneous extinction approaches zero, and thus the probability to have a resistant mutant stabilizes at a constant level.

The quantities P_{ext} and $P_{\text{non-prod}}$ are related to the probabilities of treatment success and failure. The probability of treatment failure can be defined as

$$P_{\text{failure}}(t) = 1 - \xi_0^{M_0}.$$

To define the probability of treatment success, we note that the quantity $\varphi_{0,\ldots,0}(t) = \Psi(0, \ldots, 0; t)$ is the probability of having zero cells of all types at time t. This probability includes (i) the scenario where the colony goes extinct spontaneously, and (ii) the scenario where the tumor grows and is subsequently treated successfully. The latter process has the meaning of the probability of treatment success. For small mutation rates, the probability of scenario (i) can be approximated as $(D_0/L_0)^{M_0}$. We have,

$$\varphi_{0,\ldots,0}(t) = (D_0/L_0)^{M_0} + (1 - (D_0/L_0)^{M_0})P_{\text{success}}(t),$$

where M_0 denotes the initial number of wild-type cells, and L_0 an D_0 are division and death rates of susceptible (wild-type) cells before treatment starts. Thus we have

$$P_{\text{success}}(t) = \frac{\xi_0^{M_0}(t) - (D_0/L_0)^{M_0}}{1 - (D_0/L_0)^{M_0}}. \tag{4.10}$$

4.2.6 Symmetric Coefficients

All the resistant types can be separated into classes such that in each class k, all the types are resistant to exactly k drugs (and susceptible to $m - k$ drugs). For each k, the class consists of all variables ξ_s such that the binary numbers s contain exactly k nonzero entries. Therefore, we can denote by ξ_k, with $0 \leq k \leq m$, the class of variables describing resistance to k drugs. Let us suppose that within each class, the birth and death rates are equal, and also that all mutation rates are equal to each other. Because of this symmetry assumption, it does not matter which k drugs we pick. The total number of distinct equations in this case is not $n = 2^m + 1$, but $m + 1$:

$$\dot{\xi}_k = L_k(1 - (m - k)u)\xi_k^2 + [(m - k)L_k u\xi_{k+1} - (L_s + d_k)]\xi_k + d_k, \quad 0 \leq k \leq m.$$

(4.11)

We can further simplify the description by assuming a fully symmetrical case, where all the rate coefficients are the same for all types and all drugs.

4.3 Example: The Case of Two Drugs

Let us suppose that the treatment is a combination of $m = 2$ drugs. Therefore we can distinguish four phenotypes, "A^{00}", "A^{10}", "A^{01}" and "A^{11}". The Kolmogorov forward equation is givenby:

$$\dot{\varphi}_{i_{00},i_{10},i_{01},i_{11}} = \left[\varphi_{i_{00}-1,i_{10},i_{01},i_{11}} L_{00}(i_{00} - 1)(1 - u_1 - u_2 - u_{12}) + \right.$$

$$\varphi_{i_{00},i_{10}-1,i_{01},i_{11}} L_{10}(i_{10} - 1)(1 - u_2 - u_{12}) +$$

$$\left. \varphi_{i_{00},i_{10},i_{01}-1,i_{11}} L_{01}(i_{01} - 1)(1 - u_1 - u_{12}) + \varphi_{i_{00},i_{10},i_{01},i_{11}-1} L_{10}(i_{11} - 1) \right] +$$

$$\left[i_{00}(u_1 \varphi_{i_{00},i_{10}-1,i_{01},i_{11}} + u_2 \varphi_{i_{00},i_{10},i_{01}-1,i_{11}} + u_{12} \varphi_{i_{00},i_{10},i_{01},i_{11}-1}) \right.$$

$$+ i_{10}(u_2 \varphi_{i_{00},i_{10},i_{01},i_{11}-1} + u_{12} \varphi_{i_{00},i_{10},i_{01},i_{11}-1})$$

$$\left. + i_{01}(u_1 \varphi_{i_{00},i_{10},i_{01},i_{11}-1} + u_{12} \varphi_{i_{00},i_{10},i_{01},i_{11}-1}) \right] +$$

$$\left[d_{00} \varphi_{i_{00}+1,i_{10},i_{01},i_{11}} + d_{10} \varphi_{i_{00},i_{10}+1,i_{01},i_{11}} + \right.$$

$$\left. d_{01} \varphi_{i_{00},i_{10},i_{01}+1,i_{11}} + d_{11} \varphi_{i_{00},i_{10},i_{01},i_{11}+1} \right] -$$

$$\varphi_{i_{00},i_{10},i_{01},i_{11}}\Big[i_{00}(L_{00}+d_{00}) + i_{10}(L_{10}+d_{10})$$

$$+ i_{01}(L_{01}+d_{01}) + i_{11}(L_{11}+d_{11})\Big]. \tag{4.12}$$

In this master equation, the first term in square brackets on the right-hand side comprises all the processes of faithful cell division, the second term in square brackets includes all the mutation events, the third one represents all the cell death events, and the fourth term corresponds to no change in the system's state.

4.3.1 Equations for the Moments

From the master equation, information about all the moments can be extracted. In particular, the equations for the mean numbers of cells in each class can be written. Let us denote

$$x_{00}(t) = \sum \varphi_{i_{00},i_{10},i_{01},i_{11}}(t)i_{00}, \quad x_{10}(t) = \sum \varphi_{i_{00},i_{10},i_{01},i_{11}}(t)i_{10},$$

$$x_{01}(t) = \sum \varphi_{i_{00},i_{10},i_{01},i_{11}}(t)i_{01}, \quad x_{11}(t) = \sum \varphi_{i_{00},i_{10},i_{01},i_{11}}(t)i_{11},$$

where the summation is performed over all the four indices. Then we have:

$$\dot{x}_{00} = [L_{00}(1-u_1-u_2)-d_{00}]x_{00}, \tag{4.13}$$

$$\dot{x}_{10} = [L_{10}(1-u_2)-d_{10}]x_{10} + L_{00}u_1x_{00}, \tag{4.14}$$

$$\dot{x}_{01} = [L_{01}(1-u_1)-d_{01}]x_{01} + L_{00}u_2x_{00}, \tag{4.15}$$

$$\dot{x}_{11} = [L_{11}-d_{11}]x_{11} + L_{10}u_2x_{10} + L_{01}u_1x_{01}. \tag{4.16}$$

These equations can be obtained directly from the master equation; for example, the first equation is nothing but equation (4.12) multiplied by i_{00} and summed over all the indices. The initial conditions can be written as

$$x_{00}(0) = M_0, \quad x_{10}(0) = x_{01}(0) = x_{11}(0) = 0. \tag{4.17}$$

In other words, we assume that at time zero, there are M_0 fully susceptible cells, and no mutants are initially present. The deterministic equations obtained in this way can help one reason about the expected dynamic of the colony growth and resistance generation. Equations of this type are used, for example, in Chaps. 7 and 9. However, they cannot address questions of the probability of treatment success. To quantify the likelihood of a successful treatment outcome, we need to use the stochastic approach, which is described next.

4.3.2 Equations for the Characteristics

The probability generating function is defined in accordance with equation (4.5),

$$\Psi(\xi_{00}, \xi_{10}, \xi_{01}, \xi_{11}; t) = \sum_{s=0}^{n} \varphi_{i_{00}, i_{10}, i_{10}, i_{11}}(t) \xi_{00}^{i_{00}} \xi_{10}^{i_{10}} \xi_{01}^{i_{01}} \xi_{11}^{i_{11}}.$$

The equations for the characteristics are as follows:

$$\dot{\xi}_{11} = L_{11}\xi_{11}^2 - (L_{11} + d_{11})\xi_{11} + d_{11}, \tag{4.18}$$

$$\dot{\xi}_{10} = L_{10}(1 - u_2)\xi_{10}^2 + [L_{10}u_2\xi_{11} - (L_{10} + d_{10})]\xi_{10} + d_{10}, \tag{4.19}$$

$$\dot{\xi}_{01} = L_{01}(1 - u_1)\xi_{01}^2 + [L_{01}u_1\xi_{11} - (L_{01} + d_{01})]\xi_{01} + d_{01}, \tag{4.20}$$

$$\dot{\xi}_{00} = L_{00}(1 - u_1 - u_2)\xi_{00}^2 + [L_{00}(u_1\xi_{10} + u_2\xi_{01}) - (L_{00} + d_{00})]\xi_{00} + d_{00}, \tag{4.21}$$

with general initial conditions (4.8). Solutions of this system with the initial conditions

$$\xi_{00} = \xi_{10} = \xi_{01} = \xi_{11} = 0 \tag{4.22}$$

are used to calculate the probability of treatment success. Eqs. (4.18–4.21) can be solved recursively. Let us assume the absence of treatment ($d_s = D_s$). We make the change of variables,

$$\xi_s = -\frac{\dot{X}_s}{L_s \left(1 - u^{s,\text{out}}\right) X_s}, \quad 0 \le s \le n \tag{4.23}$$

and obtain a Riccatti-type equation for X_s:

$$\ddot{X}_s + \left[L_s \left(1 - \sum_j u_j^{s,\text{out}}\xi_j\right) + D_s\right] \dot{X}_s + L_s \left(1 - u^{s,\text{out}}\right) D_s X_s = 0. \tag{4.24}$$

Note that the solution, X_s, depends of the functions ξ_j, the variables downstream from the node s. For $D_s \ne 0$, only the first of the equations, the one for X_{11}, can be solved analytically because there are no variables downstream from node (11), and we have a Riccatti equation with constant coefficients. All the rest of the equations in general have to be solved numerically.

One way to get some analytical insights is to suppress the dynamics of type "A^{11}", such that the equation for ξ_{11} becomes

$$\dot{\xi}_{11} = 0. \tag{4.25}$$

In this case the quantity

$$1 - \xi_s(t)$$

stands for the probability to create at least one mutant of type "A^s" by time t; note that this mutant may have died away by time t, or it may have created offspring: because of equation (4.25) we do not distinguish between these scenarios. From initial conditions (4.22) we have $\xi_{11}(t) = 0$, and the two equations for the variables corresponding to resistance to one drug can be solved analytically. These variables correspond to the indices s that contain only one nonzero entry. We have,

$$\xi_{01} = -\frac{b_1 + Ab_2 e^{(b_2-b_1)t}}{L_{01}(1-u_1)(1 + Ae^{(b_2-b_1)t})}, \tag{4.26}$$

where

$$A = -\frac{b_1 + L_{01}(1-u_1)}{b_2 + L_{01}(1-u_1)},$$

and $b_1 > b_2$ are roots of the quadratic equation,

$$b^2 + (L_{01} + D_{01})b + L_{01}(1-u_1)D_{01} = 0. \tag{4.27}$$

Similarly, ξ_{10} is obtained by changing $u_1 \to u_2$, $L_{01} \to L_{10}$, and $D_{01} \to D_{10}$.

To get the answer for ξ_{00}, these expressions should be substituted in Eq. (4.21) to continue the recursion. Only a solution by numerical integration is possible, unless further simplifying assumptions are made.

4.4 Mutant Production Before and During Treatment

4.4.1 General Theory

At what stage are mutants predominantly generated: before the start of treatment, or after therapy is applied? Let us introduce two quantities, $P^{\uparrow}(N)$ and $P^{\downarrow}(N)$, in order to be able to characterize and compare the production of mutants before therapy and during therapy.

To simplify the description, we will assume the symmetry in coefficients described in Sect. 4.2.6, so that the number of equations for m drugs is $m + 1$.

Preexistence of resistant mutants. Let us suppose that we can switch off the mutation rate during therapy, and calculate the probability of nonextinction (which is the same as the probability to have resistant mutants as $t \to \infty$). Let us find the probability to have mutants in the limit when $t \to \infty$. We have the following systems of equations. Before start of treatment,

$$\dot{\xi}_m = L\xi_m^2 - (L + D)\xi_m + D, \tag{4.28}$$

$$\dot{\xi}_i = L(1 - lu)\xi_i^2 + (lLu\xi_{i+1} - (L + D))\xi_i + D, \quad 0 \le i < m, \tag{4.29}$$

and upon the start of therapy,

$$\dot{\xi}_m = L\xi_m^2 - (L + D)\xi_m + D, \tag{4.30}$$

$$\dot{\xi}_i = L(1 - iu)\xi_i^2 - (L + D + H)\xi_i + D + H. \quad 0 \le i < m, \tag{4.31}$$

In order to solve this general problem, we need to implement the method described in Sect. 4.2.4. We are interested in the quantity

$$P^\uparrow(N) = \lim_{\bar{t} \to \infty} \Psi(0, 1, \ldots, 1, t_* + \bar{t}).$$

The meaning of $P^\uparrow$ is the probability to develop resistance, as $t \to \infty$, if mutations only happen before the start of treatment.

First we solve system (4.30–4.31) and find its limiting behavior. Then we use the steady-state values for all the variables as the initial condition for system (4.28–4.29). The quantity describing preexistence of resistant mutants can be calculated as

$$P^\uparrow(N) = 1 - \xi_0(t_*)^{M_0}, \quad t_* = \frac{1}{L - D}\ln(N/M_0).$$

This corresponds to the probability to have at least one fully resistant mutant as $t \to \infty$, given that the size of the colony at the start of treatment is N, and no further mutants are produced during treatment.

Generation of mutants during treatment. In order to characterize the role of the treatment phase in the generation of resistance, we will switch off the mutation rates during the growth (pretreatment) phase, and turn them back on once treatment starts. The simplest way in which the problem can be formulated is as follows. We consider only the treatment phase, and start from N susceptible cells (no preexistence). We have

$$\dot{\xi}_m = L\xi_m^2 - (L + D)\xi_m + D, \tag{4.32}$$

$$\dot{\xi}_i = L(1 - iu)\xi_i^2 + [iLu\xi_{i+1} - (L + D + H)]\xi_i + (D + H), \quad 0 \le i < m, \tag{4.33}$$

$$\xi_i(0) = 0 \quad 0 \le i \le m. \tag{4.34}$$

The quantity characterizing the generation of mutants during therapy is given by

$$P^\downarrow(N) = 1 - \lim_{t \to \infty} \xi_0(t)^N.$$

This function has the meaning of the probability to have created viable mutants in the course of therapy, starting from N fully susceptible cells.

We can calculate the limiting behavior of solutions of system (4.32–4.34), as time goes to infinity, by using the method described in Sect. 4.2.5. We have, under the assumption that $u \ll (L - D)$, that

$$\lim_{t \to \infty} \xi_m = \frac{D}{L}, \tag{4.35}$$

$$\lim_{t \to \infty} \xi_i = 1 - \frac{i!(L - D)L^{i-1}u^i}{(D + H - L)^i}, \quad 0 \le i < m. \tag{4.36}$$

If the total number of drugs used is m, then the probability of mutant generation with no preexistence is given by

$$P^{\downarrow}(N) = 1 - \left(1 - \frac{m!(L - D)L^{m-1}u^m}{(D + H - L)^m}\right)^N. \tag{4.37}$$

We observe that the larger the initial size, the lower is the probability of treatment success.

In the rest of this section we will calculate and compare the quantities $P^{\uparrow}$ and $P^{\downarrow}$. If it turns out that $P^{\uparrow}(N) > P^{\downarrow}(N)$, then we can conclude that resistance is generated at a higher intensity before therapy, and the contribution of the therapy phase is less important. The opposite result, $P^{\uparrow}(N) < P^{\downarrow}(N)$, would tell us that most mutants are generated after therapy begins. The biological consequences of the results are discussed in more detail in the next chapter.

4.4.2 The Case of One Drug

Mutant production during treatment. The expression for the probability to create resistance during therapy is obtained from equation (4.37):

$$P^{\downarrow}(N) = 1 - \left(1 - \frac{(L - D)u}{H - (L - D)}\right)^N. \tag{4.38}$$

In the analysis below it is convenient to use the following threshold value of treatment intensity, H:

$$H_c = 2(L - D). \tag{4.39}$$

For $H = H_c$, we have

$$P^{\downarrow}(N) = 1 - (1 - u)^N \approx 1 - e^{-Nu}. \tag{4.40}$$

For values of H larger than the threshold, $H > H_c$, the corresponding probability to create resistance is lower.

Mutant production before treatment. In the case of one drug, we have only two equations in system (4.28–4.29). The equation for $\xi_m = \xi_1$ can be solved exactly. However, the equations for ξ_0 will contain nonconstant coefficients and cannot be solved analytically (unless $D = 0$). Therefore, instead of solving systems (4.28–4.29, 4.30–4.33) directly, we will use a different method.

Let us first suppress the dynamics of the resistant mutant ($\dot{\xi}_1 = 0$) and calculate the probability of mutant generation during the pretreatment phase. We have $\xi_1 = 0$, and thus the equation for ξ_0 reads

$$\dot{\xi}_0 = L(1-u)\xi_0^2 - (L+D)\xi_0 + D.$$

This Riccatti equation can be solved to yield

$$\xi_0 = -\frac{\beta_1 + A\beta_2 e^{(\beta_2-\beta_1)t}}{a(1 + Ae^{(\beta_2-\beta_1)t})},$$

with

$$\beta_{1,2} = \frac{1}{2}\left(-(L+D) \pm \sqrt{(L+D)^2 - 4LD(1-u)}\right),$$

and

$$A = -\frac{\beta_1 + L(1-u)}{\beta_2 + L(1-u)}.$$

Taking the highest order terms in u, we obtain,

$$\xi_0(t) = 1 - \frac{(e^{(L-D)t} - 1)Lu}{L - D} \approx 1 - \frac{NLu}{M_0(L-D)}, \tag{4.41}$$

and the probability to have created a mutant is simply

$$P_{\text{create}} = 1 - \xi_0^{M_0} \approx \frac{LuN}{L-D}.$$

This quantity has the meaning of the probability to have produced at least one resistant mutant by the time the colony has reached size N (this mutant and all its progeny may or may not be present at this point).

This is a very intuitive result, as the probability to have created a mutant is given by the total number of cell divisions, $\mathcal{N}$, times the mutation rate, u. The total number of cell divisions from one cell to N cells is roughly given by $\mathcal{N} = NL/(L - D)$, which results in the formula for P_{create} above.

Now, each resistant mutant created during the pretreatment phase, will give rise to a lineage of progeny, with the growth rate L and the death rate D, which is not affected by the presence of the drug. The probability for a lineage starting from one cell to survive (not to go extinct as $t \to \infty$) is given by $P_{\text{survive}} = 1 - D/L$. Therefore, the probability to generate resistance during the pretreatment phase is given by

$$P^{\uparrow}(N) = P_{\text{create}} P_{\text{survive}} = Nu.$$

This result was obtained by intuitive reasoning, and it obviously holds only when $Nu < 1$. A more rigorous derivation for the quantity $P^{\uparrow}$ is possible, and is presented next.

Let us consider one equation,

$$\dot{x}_1 = Lx_1^2 - (L + D)x_1 + D, \quad x_1(0) = 1,$$

Here, the quantity $1 - x_1(t)$ has the meaning of the probability that a one-hit colony survives until time t. We have,

$$1 - x_1(t) = \frac{L - D}{L - De^{-(L-D)t}}. \tag{4.42}$$

Using doubly stochastic process, we can calculate the probability that the colony has at least one mutant by time t,

$$1 - exp\left\{-uL \int_0^t N(t')[1 - x_1(t - t')] \, dt'\right\},$$

where the multiplier $uN(t')$ reflects the production of mutants and the quantity $(1 - x_1)$ corresponds to the survival probability of each mutant lineage. We set $N(t) = M_0 e^{(L-D)t}$. Integration gives

$$1 - exp\left\{-Nu\frac{L}{D}\left(Dt + \ln\frac{Le^{-Dt} - De^{-Lt}}{L - D}\right)\right\}. \tag{4.43}$$

This quantity corresponds to the probability to have at least one resistant mutant by the time when therapy starts. Note that in the limit $D \to 0$ we obtain

$$P^{\uparrow} = 1 - e^{-u(N(t)-M_0)},$$

which coincides exactly with the result by [35], eq. (4). Equation (4.43) is a generalization of the formula by [35] for nonzero death rates.

In order to calculate the quantity $P^{\uparrow}(N)$, we need to find the limiting behavior of the probability to have a mutant as $t \to \infty$, and also to ignore the production of new mutants during the treatment stage. Instead of the time-dependent formula (4.42) we can use its limiting value at $t \to \infty$, $1 - D/L$, so we have

$$P^{\uparrow}(N) = 1 - exp\left\{-M_0 u L e^{(L-D)t} \int_0^t e^{-(L-D)t'}\left(1 - \frac{D}{L}\right) dt'\right\} = 1 - e^{-Nu}.$$
(4.44)

We can see that this is exactly equal to expression (4.40).

4.4.3 The Case of Two Drugs

Mutant production during treatment. From expression (4.37) we obtain in the case $m = 2$,

$$P^{\downarrow}(N) = 1 - \left(1 - \frac{2(L-D)Lu^2}{(H-(L-D))^2}\right)^N.$$

Although Nu may or may not be a small quantity, it is safe to assume that $Nu^2 \ll 1$, which allows the following approximation:

$$P^{\downarrow}(N) = \frac{2(L-D)Lu^2 N}{(H-(L-D))^2}.$$
(4.45)

Mutant production before treatment. In the case of two drugs, let us calculate the probability of double-mutant creation as a function of the tumor size. Again, we will suppress the dynamics of the double mutants, such that $\xi_0 \equiv 0$, and only keep track of the creation process. We will make the approximation of a doubly stochastic process, whereby generation of each one-hit mutant leads to a birth–death process, all of which are independent and identically distributed (see also [36, 37]). We will assume that the total population size changes according to the deterministic exponential law, $N(t) = M_0 e^{(L-D)t}$ (see [34] for extended discussion). Generalizing the notion of a filtered Poisson processes described, e.g., in [38], we have,

$$\Psi(0, 1; t) = exp\left[-2Lu \int_0^t N(t')\left(1 - \Phi(0, 1; t - t')\right) dt'\right],$$
(4.46)

where $LuN(t')dt'$ is the probability to create a one-hit mutant in the time-interval $(t, t + dt)$, and $1 - \Phi(0, 1; t - t')$ is the probability that the lineage resulting from that mutant will give rise to the production of a double-mutant in the time from the creation of the lineage, t', to the current time, t. The latter probability is given by $1 - \xi_1(t - t')$, see formula (4.41). The factor 2 in the exponent comes from the two possibilities of acquiring two hits. We have,

$$P_{create} = 2\left(\frac{Lu}{L-D}\right)^2 N\left(\ln\frac{N}{M_0} - 1\right).$$
(4.47)

The method of a doubly-stochastic process is a good approximation as long as the one-hit mutants are "rare". Multiplying P_{create} by the probability of each double-mutant to survive, $P_{\text{survive}} = 1 - D/L$, we obtain

$$P^{\uparrow}(N) = \frac{2Lu^2N}{L-D}\left(\ln\frac{N}{M_0} - 1\right),\qquad(4.48)$$

that is, this quantity now depends on the turnover rate, D/L. Note that formula (4.48) breaks down as $D \to L$ as the doubly stochastic approximation is not applicable in this regime anymore; at this moment we do not have a method to handle the regime $D \approx L$.

4.5 Probability of Treatment Success

4.5.1 Analytical Results

Let us solve the two-stage problem for the case of one drug, $m = 1$. Therapy starts at the time, t_* (and the colony size reaches the value N). Before start of treatment, we have system (4.28–4.29), and during treatment, we have equations (4.32–4.33).

In order to solve this problem, let us use the method of Sect. 4.2.4. We first need to find the limiting values of ξ_i under the treatment conditions (system (4.32–4.33)), as given by formulas (4.35–4.36), and use these as the initial conditions for equations (4.28–4.29), in the interval $0 \le t \le t_*$, where t_* is the time when treatment starts. The quantities $[\xi_i(t_*)]^{M_0}$ are the probabilities of treatment success with therapy starting at time t_*.

Equation (4.28) for $\xi_1(t)$ with initial condition (4.35) can be solved exactly to give

$$\xi_1(t) = \frac{D}{L}$$

The equation for ξ_0 is a constant coefficient Riccatti equation,

$$\dot{\xi}_0 = L(1-u)\xi_0^2 + (Du - (L+D))\xi_0 + D,\quad \xi_0(0) = 1 - \frac{(L-D)u}{D+H+L}.$$

We can write down the exact solution:

$$\xi_0 = \frac{Ae^{(D[1-u]-L)t}L + D(1-u)}{L(1-u)[1 + Ae^{(D[1-u]-L)t}]},\quad A = \frac{(1+u)(L-D)(D+H-L[1+u])}{Lu[H + (D-L)u]}.$$

The limiting behavior is given by

$$\lim_{t_* \to \infty} \xi_0 = \frac{D}{L}.$$

In the case where $Nu \lesssim 1$, we can take the limit $u \to 0$ and find an approximate formula for $\xi_0(t)$,

$$\xi_0 \approx 1 - \left(\frac{He^{(L-D)t}}{H+D-L} - 1 \right) u.$$

Setting

$$e^{(L-D)t} = \frac{N}{M_0},$$

and neglecting 1 compared to this quantity, we can write down the probability of treatment success,

$$P_{\text{success}}(N) \approx \left(1 - \frac{HNu}{(H+D-L)M_0} \right)^{M_0}. \tag{4.49}$$

We can see that larger values of D correspond to a higher probability of treatment success. If therapy is very strong such that $H \gg L - D$, we have a very simple formula,

$$P_{\text{success}}(N) \approx \left(1 - \frac{N}{M_0}u \right)^{M_0},$$

i.e., treatment success does not depend on D for strong therapies.

4.5.2 Threshold Tumor Size

A convenient measure of resistance is the size, N_δ, corresponding to the probability of treatment failure equal to a fixed amount, $\delta \ll 1$. We will call the quantity N_δ the *threshold tumor size*. This quantity is used later in this book to characterize treatment success. In the case of one drug, the value of N_δ can be calculated. We have

$$\frac{HN_\delta u}{(H+D-L)} = \delta,$$

and therefore

$$N_\delta = \frac{1}{u}\frac{(H+D-L)\delta}{H},$$

that is, the log size is a linear function of $|\ln u|$ with slope $n = 1$. Also, we can see that for strong therapies ($H \gg (L - D)$), the value of N_δ does not depend on the turnover rate, D/L.

Problems

4.1. Derive the partial differential Eq. (4.2) from the Kolmogorov forward Eq. (4.1), in the case of a one-drug treatment.

4.2. The same in the case of 2-drug treatments: Derive the partial differential equation and the equations for characteristics, (4.13–4.16), from Kolmogorov forward Eq. (4.12). How does the description simplify in the case of symmetric coefficients, $L_s = L$, $D_s = D$, etc?

4.3. In the general case of m-drug treatments: Derive the partial differential Eq. (4.6) from the Kolmogorov forward Eq. (4.4). Show that the rewrite rules listed in Sect. 4.2.3 hold.

4.4. Apply the method of characteristics outlined in Sect. 4.2.4 to the PDE in the case of one-drug treatment, (4.2). Find the probability of treatment success assuming that at $t = 0$, there are M_0 susceptible mutants, and treatment starts at time t_*.

4.5. Show that the probability of spontaneous colony extinction starting from M_0 cells in the absence of treatment is given by $(D_0/L_0)^{M_0}$, where L_0 and D_0 are division and death rates of the cells. This result was used in the derivation of formula (4.10). (*Hint*: this is similar to the gambler's ruin problem).

4.6. Research project
Determining the number of mutations in a growing colony of cells is related to the famous Luria-Delbrook distribution. Find out about the history and major benchmarks in the efforts to solve this problem.

4.7. Research project
In this book, approximation (4.9) was used, which postulates a deterministic connection between the total cell number and the time elapsed. What are the consequences of this assumption (see ref. [34])?

4.8. Numerical project
Use any programming language to set up a simulation that calculates the probability of treatment success, given a treatment with m drugs that begins at time t_* (with approximation (4.9)). Assume that at $t = 0$, the colony consists of M_0 susceptible mutants. First solve the system for the characteristic equations under the treatment conditions (for infinitely long treatment, the result can be found from the stable equilibrium of the system for characteristics). Then use this result as the initial condition for the system with pretreatment coefficients.

References

1. Goldie, J.H., Coldman, A.J.: A mathematic model for relating the drug sensitivity of tumors to their spontaneous mutation rate. Cancer Treat. Rep. **63**(11–12), 1727–1733 (1979)
2. Goldie, J.H., Coldman, A.J.: A model for resistance of tumor cells to cancer chemotherapeutic agents. Math. Biosci. **65**, 291–307 (1983)

3. Goldie, J.H., Coldman, A.J.: Quantitative model for multiple levels of drug resistance in clinical tumors. Cancer Treat. Rep. **67**(10), 923–931 (1983)
4. Coldman, A.J., Goldie, J.H.: Role of mathematical modeling in protocol formulation in cancer chemotherapy. Cancer Treat. Rep. **69**(10), 1041–1048 (1985)
5. Goldie, J., Coldman, A.: A model for tumor response to chemotherapy: an integration of the stem cell and somatic mutation hypotheses. Cancer Invest. **3**(6), 553–564 (1985)
6. Coldman, A., Goldie, J.: A stochastic model for the origin and treatment of tumors containing drug-resistant cells. Bull. Math. Biol. **48**(3), 279–292 (1986)
7. Goldie, J.H., Coldman, A.J.: Drug Resistance in Cancer: Mechanisms and Models. Cambridge University Press, Cambridge (1998)
8. Kimmel, M., Axelrod, D.E.: Mathematical models of gene amplification with applications to cellular drug resistance and tumorigenicity. Genetics **125**(3), 633–644 (1990)
9. Harnevo, L.E., Agur, Z.: The dynamics of gene amplification described as a multitype compartmental model and as a branching process. Math. Biosci. **103**(1), 115–138 (1991)
10. Harnevo, L.E., Agur, Z.: Use of mathematical models for understanding the dynamics of gene amplification. Mutat. Res. **292**(1), 17–24 (1993)
11. Axelrod, D.E., Baggerly, K.A., Kimmel, M.: Gene amplification by unequal sister chromatid exchange: probabilistic modeling and analysis of drug resistance data. J. Theor. Biol. **168**(2), 151–159 (1994)
12. Kimmel, M., Stivers, D.N.: Time-continuous branching walk models of unstable gene amplification. Bull. Math. Biol. **56**(2), 337–357 (1994)
13. Cojocaru, L., Agur, Z.: A theoretical analysis of interval drug dosing for cell-cycle-phase-specific drugs. Math. Biosci. **109**(1), 85–97 (1992)
14. Kimmel, M., Swierniak, A., Polanski, A.: Infinite-dimensional model of evolution of drug resistance of cancer cells. J. Math. Syst. Est. Contr. **8**, 1–16 (1998)
15. Coldman, A.J., Murray, J.M.: Optimal control for a stochastic model of cancer chemotherapy. Math. Biosci. **168**(2), 187–200 (2000)
16. Swierniak, A., Smieja, J.: Cancer chemotherapy optimization under evolving drug resistance. Nonlin. Anal. **47**, 375–386 (2001)
17. Murray, J.M., Coldman, A.J.: The effect of heterogeneity on optimal regimens in cancer chemotherapy. Math. Biosci. **185**(1), 73–87 (2003)
18. Smieja, J., Swierniak, A.: Different models of chemotherapy taking into account drug resistance stemming from gene amplification. Int. J. Appl. Math. Comput. Sci. **13**(3), 297–305 (2003)
19. Swan, G.W.: Role of optimal control theory in cancer chemotherapy. Math. Biosci. **101**(2), 237–284 (1990)
20. Gaffney, E.A.: The application of mathematical modelling to aspects of adjuvant chemotherapy scheduling. J. Math. Biol. **48**(4), 375–422 (2004)
21. Gaffney, E.A.: The mathematical modelling of adjuvant chemotherapy scheduling: incorporating the effects of protocol rest phases and pharmacokinetics. Bull. Math. Biol. **67**, 563–611 (2005)
22. Gardner, S.N., Fernandes, M.: New tools for cancer chemotherapy: computational assistance for tailoring treatments. Mol. Cancer Ther. **2**(10), 1079–1084 (2003)
23. Jackson, T.L., Byrne, H.M.: A mathematical model to study the effects of drug resistance and vasculature on the response of solid tumors to chemotherapy. Math. Biosci. **164**(1), 17–38 (2000)
24. Byrne, H.M., Chaplain, M.A.: Growth of nonnecrotic tumors in the presence and absence of inhibitors. Math. Biosci. **130**(2), 151–181 (1995)
25. Kendal, W.S., Frost, P.: Pitfalls and practice of Luria-Delbruck fluctuation analysis: a review. Cancer Res. **48**(5), 1060–1065 (1988)
26. Jaffrezou, J.P., Chen, G., Duran, G.E., Kuhl, J.S., Sikic, B.I.: Mutation rates and mechanisms of resistance to etoposide determined from fluctuation analysis. J. Natl. Cancer Inst. **86**(15), 1152–1158 (1994)
27. Chen, G.K., Duran, G.E., Mangili, A., Beketic-Oreskovic, L., Sikic, B.I.: MDR 1 activation is the predominant resistance mechanism selected by vinblastine in MES-SA cells. Br. J. Cancer **83**(7), 892–898 (2000)

28. Coldman, A.J., Goldie, J.H.: A stochastic model for the origin and treatment of tumors containing drug-resistant cells. Bull. Math. Biol. **48**(3–4), 279–292 (1986)
29. Komarova, N.L., Wodarz, D.: Drug resistance in cancer: principles of emergence and prevention. Proc. Natl. Acad. Sci. U.S.A. **102**(27), 9714–9719 (2005)
30. Komarova, N.: Stochastic modeling of drug resistance in cancer. J. Theor. Biol. **239**(3), 351–366 (2006)
31. Katouli, A., Komarova, N.: Optimizing combination therapies with existing and future cml drugs. PLoS ONE **5**(8), 300 (2010) (e12)
32. Karlin, S., Taylor, H.E.: A First Course in Stochastic Processes. Academic press, New York (1975)
33. Gockenbach, M.S.: Partial Differential Equations: Analytical and Numerical Methods. Siam, Philadelphia (2010)
34. Komarova, N., Wu, L., Baldi, P.: The fixed-size luria-delbruck model with a nonzero death rate. Math. Biosci. **210**(1), 253–290 (2007)
35. Goldie, J.H., Coldman, A.J.: A model for resistance of tumor cells to cancer chemotherapeutic agents. Math. Biosci. **65**, 291–307 (1983)
36. Moolgavkar, S.H., Dewanji, A., Venzon, D.J.: A stochastic two-stage model for cancer risk assessment. i. the hazard function and the probability of tumor. Risk Anal. **8**(3), 383–392 (1988)
37. Iwasa, Y., Michor, F., Nowak, M.A.: Stochastic tunnels in evolutionary dynamics. Genetics **166**(3), 1571–1579 (2004)
38. Parzen, E.: Stochastic Processes. Holden-Day, San Francisco (1962)

Chapter 5
Evolutionary Dynamics of Drug Resistant Mutants in Targeted Treatment of CML

Abstract This chapter discusses the basic evolutionary dynamics of cells that are resistant to small molecule inhibitors in the context of an exponentially growing population of cells, as typically observed in the treatment of CML blast crisis. The basic principles of resistance emergence is discussed. Mathematical models clearly indicate that the pre-treatment tumor growth phase is crucial for the generation of resistant mutants, and that the treatment phase itself does not significantly contribute to the generation of resistant mutants. In this respect, the growth kinetics of the tumor plays an important role. Tumors characterized by a higher turnover of cells have a higher probability to harbor resistant cells at any given tumor size, compared to low-turnover tumors. These insights are then applied to calculate the probability that a tumor of a given size is simultaneously resistant to a number of m drugs that can be given in combination, based on the parameters of the system. These calculations suggest that in the absence of cross-resistance, a combination of three drugs can prevent resistance-induced treatment failure in CML blast crisis.

Keywords Treatment of CML · Resistance generation · Combinatorial mutation diagrams · Resistant cells · Susceptible cells · Evolution of resistance · Pre-treatment phase · Probability of treatment success · Probability of treatment failure · Multiple drug treatments · Combination treatments · Tumor turnover · Prevention of resistance · Tumor size · Mutation rate

5.1 Introduction

Drug resistance is a frequent clinical problem for cancer patients [1]. Many mechanisms of drug resistance have been found [2–6]. For example, drugs can be prevented from entering the cells; drugs can be pumped out of cells; they can be enzymatically inactivated; drug activity can be prevented by mutation or altered expression of the target; defects in apoptosis, senescence, and repair mechanisms

N. L. Komarova and D. Wodarz, *Targeted Cancer Treatment in Silico*,
Modeling and Simulation in Science, Engineering and Technology,
DOI: 10.1007/978-1-4614-8301-4_5, © Springer Science+Business Media New York 2014

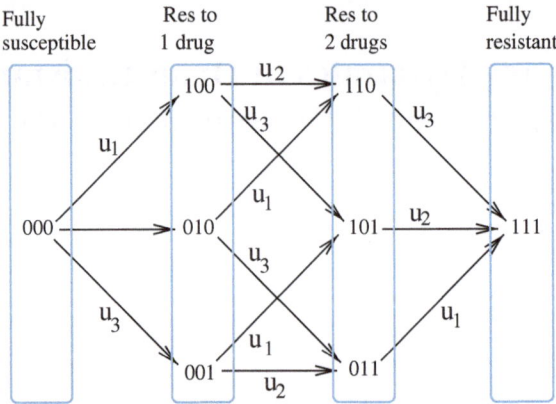

Fig. 5.1 Mutation diagram corresponding to three drugs. Each node corresponds to a phenotype. The binary numbers identify which drugs the phenotype is resistant to: e.g. 011 means that this type is resistant to drugs 2 and 3 but not to drug 1. The *leftmost* type (000) is fully susceptible, and the *rightmost* one (111) is resistant to all three drugs. The mutations rates are marked above each arrow

can contribute to resistance. A particular problem in cancer is the occurrence of multi-drug resistance. Many of these mechanisms of drug resistance apply to traditional chemotherapies. With targeted treatment using small molecule inhibitors, it is mainly point mutations, and less often, gene amplification events that confer resistance. As explained in Chap. 2, small molecule inhibitors bind to defined targets that have been shown to be crucial for the maintenance and progression of the cancer. The most prominent example currently is the treatment of CML with imatinib [7] and subsequently developed drugs that inhibit the BCR-ABL fusion gene product. Resistance typically occurs if the target site mutates, such that the drug cannot bind anymore. In the earlier phase of the disease, targeted therapy of CML is usually characterized by a successful response, resulting in a complete hematological or a complete cytogenetic remission. In the more advanced stages, however, most notably during blast crisis, drug resistance tends to be a problem and treatment success is consequently limited [8, 9]. In order to manage this problem, it is important to have an understanding of the principles that underlie the emergence of drug resistance. Mathematical models have been instrumental in examining the evolutionary dynamics of drug resistant mutants in the context of targeted therapy of CML, and this chapter explores this in the simplest scenario, i.e. in an exponentially growing population of tumor cells. This likely corresponds to advanced stages of the disease, when explosive growth occurs during blast crisis. The principles according to which drug resistant mutants are generated will be discussed, and we will explore how the combination of different drugs can potentially overcome treatment failure due to drug resistance.

5.2 The Conceptual Framework

In order to understand how resistant mutants are generated during cancer progression and treatment, we have developed the following conceptual framework (described in full detail in Chap. 4). Untreated cancerous cells are described by a stochastic birth-death process with a positive net proliferation rate. If we denote the division rate of cells as L and the death rate as D, the condition $L > D$ corresponds to a clonal expansion. We further assume that cancer is detected when the colony reaches a certain size, N, at which moment therapy starts (we will also refer to N as "treatment size"). The effect of therapy is modeled by the *drug-induced death rate*, H, which shifts the balance of birth and death such that the colony shrinks. That is, the net cell death rate of susceptible cells under treatment is larger than the division rate, $D + H > L$. If all cancerous cells were susceptible to the drug, then therapy would inevitably lead to eradication of cancer. However, in the course of cancer progression, mutations can lead to the generation of cell types, which are resistant to the drug [10]. This is assumed to occur with a probability u upon cell division. Before the tumor is treated, the mutants will behave identically to the wild-type cells. During therapy, however, the resistant phenotype will proliferate while the wild-type will be killed with a rate H. The resulting treatment failure can be countered by combining several drugs, as demonstrated effectively with viral infections [11]. Such combination therapy is included in this framework.

In our first model we assume that a mutation which confers resistance to one drug does not confer resistance to any of the other drugs in use. This may not be the case for all drugs/mutations, and these effects have to be accounted for in further modifications of the model, see Chap. 8, which studies the effects of cross-resistance. In the absence of cross-resistance, in order to become resistant to m drugs, the cell has to accumulate m mutations [10]. These mutational processes can be presented as a combinatorial mutation network, an example of which is presented in Fig. 5.1. For simplicity, we assume that mutant cells which are not resistant to all drugs in use are killed with the same rate as wild-type cells. Alternatively, it can be assumed that such mutants are partially resistant (i.e. are affected less than the wild-type but more than the fully resistant phenotype [12, 13]), but it turns out that this complication does not alter our results significantly (see below).

We will explore the principles according to which resistant mutants are generated during the pre-treatment growth phase and during therapy. In particular we investigate the chances that resistant mutants pre-exist before treating a tumor of size N. In this respect, it is key to examine the number of cell divisions which occur during the growth phase until size N is reached. This is roughly given by $\mathcal{N} = NL/(L - D)$. We can see that if $D = 0$ or $D \ll L$, the number of cell divisions is approximately given by $\mathcal{N} \approx N$. On the other hand, if D is close to L ($D \approx L$), many more cell divisions are required to reach size N, since a high death rate cancels the effect of cell divisions. For convenience, we will call the scenario where $D \approx L$ a *high-turnover* cancer. In contrast, we will call the scenario where $D = 0$ or $D \ll L$ a *low-turnover* cancer. In the following, we will first examine the emergence of resistance against a single drug and then expand the analysis to include the use of more than one drug.

5.3 Evolution of Resistance Against a Single Drug

Consider the use of one drug only. We determine the relative roles of the pre-treatment and the treatment phase for the generation of resistant mutants. In other words, how important is the pre-existence of mutants? We first perform in silico experiments where we artificially set the mutation rate to zero before nd after treatment, and calculate the corresponding probabilities of treatment failure.

First, we set the mutation rate to zero right after treatment starts. That is, mutations can only be generated before therapy. We calculate the probability of treatment failure in this setting which we define as the probability that the cancer escapes therapy due to generation of resistant mutants. This is denoted by $P^\uparrow$; the symbol $\uparrow$ indicates that we look at the contribution of the growth phase to mutant generation. We have (formula (4.44)):

$$P^\uparrow = 1 - e^{-Nu}.$$

This result means that in the context of a single drug, high-turnover and low-turnover cancers behave in exactly the same way, as far as the pre-existence of mutants is concerned. An intuitive explanation is as follows. A higher-turnover cancer requires more cell divisions to reach size N, and thus more mutants are created. At the same time, however, the death rate of the mutants is also increased. The two effects cancel each other out. Similar behavior was observed numerically in [14].

Next, we set the mutation rate to zero in the *pre-treatment* phase. Now, mutations can only be generated during therapy, and we can evaluate the corresponding contribution to treatment failure, $P^\downarrow$. From formula (4.38) we have:

$$P^\uparrow = 1 - \left(1 - \frac{u}{H/(L-D)-1}\right)^N \approx 1 - e^{-\frac{Nu}{H/(L-d)-1}}. \tag{5.1}$$

We can see that when $H = H_c \equiv 2(L - D)$, the equation for $P^\uparrow$ is the same as the equation for $P^\downarrow$. For $H > H_c$ we have

$$\frac{P^\downarrow}{P^\uparrow} < 1.$$

That is, as long as $H > H_c$, the generation of resistant mutants takes place predominantly *before* treatment starts. We argue that the treatment phase only becomes important for the generation of resistance in the unrealistic case where $H \leq H_c$. Under this condition, treatment is very ineffective such that the number of cell divisions during treatment is higher than the number of cell divisions during the growth

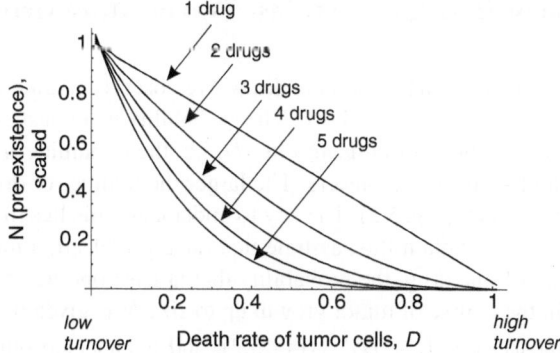

Fig. 5.2 Production of resistant mutants before treatment, depending on the death rate of tumor cells, D. To consider the pre-treatment phase only, we artificially set the mutation rate to zero upon start of therapy. Further, to concentrate on the *production* of resistant mutants, we suppressed the dynamics of the fully-resistant mutants. We plot the tumor size, N, at which the probability of treatment failure due to pre-existence equals δ. Note that all *curves* are scaled to be displayed on one graph. For a single drug, this dependence is linear. For larger number of drugs, this dependence becomes increasingly stronger than linear. Parameter values were chosen as follows: $L = 1$, $u = 10^{-6}$, $\delta = 0.01$

phase before treatment. In other words, the time it takes to eradicate the tumor by drugs in the absence of resistance is larger than the age of the tumor upon start of therapy. This is not a likely scenario. Therefore, we conclude that for all realistic cases, resistance develops before the start of treatment.

To summarize so far, two important messages follow from our theory:

- No matter what the birth-to-death ratio is for the cells, the probability of resistance generation before treatment in the case of one drug is given by $1 - e^{Nu}$. That is, it does not depend of the turn-over rate of the colony.
- For realistic treatment intensity values ($H > H_c$), we have $P^\uparrow > P^\downarrow$, that is, resistance is mostly generated before the start of therapy.

Combining the growth and the treatment phase, we can calculate the overall probability of treatment failure as a function of treatment size, N. We have from formula (4.49) with $M_0 = 1$ and $H \gg L - D$,

$$P_{\text{failure}} \approx P^\uparrow = 1 - e^{-Nu}. \tag{5.2}$$

That is, in the case of strong treatments, the outcome of one-drug therapy is independent of the turnover of the cancer, and is mostly determined by the pre-existence of resistant mutants.

5.4 Evolution of Resistance Against Two or More Drugs

Now, consider treatment with two or more drugs. We observe an important difference compared to the one drug scenario above: The probability that resistant mutants pre-exist now depends on the natural death rate, D, i.e. the dynamics are different for high-turnover and low-turnover cancers. The larger the number of drugs in use, the stronger this dependency (Fig. 5.2). The key to understand this lies in the process of mutant *generation*. To explain this, assume that once produced, a mutant does not die. In the context of one drug, the probability that at least one resistant mutant has been produced in the course of tumor growth up to size N is given by $u\mathcal{N} = \frac{NLu}{L-D}$. This depends linearly on $(L - D)^{-1}$ (Fig. 5.2), and is canceled out by the factor $(1 - D/L)$ if mutant death is included. For two drugs (requiring a double mutant), this probability is roughly given by (see formula 4.47)

$$2 \left(\frac{Lu}{L - D} \right)^2 N \ln N.$$

The dependence on $(L - D)^{-1}$ is now stronger than linear and is not canceled out anymore if mutant death is included. In general, if the number of drugs is increased, a higher natural death rate of tumor cells, D, contributes increasingly to the production of resistant mutants and thus to treatment failure (Fig. 5.2).

Another difference compared to the one drug scenario is that with higher numbers of drugs, the treatment phase becomes completely insignificant in the context of resistance generation. For instance, for $m = 2$ drugs, the expressions for $P^\uparrow$ and $P^\downarrow$ are given respectively by equations (4.48) and (4.45), and we have

$$\frac{P^\downarrow}{P^\uparrow} = (H/(L - D) - 1)^{-2}(\ln N - 1)^{-1}.$$

It is easy to see that the condition

$$\frac{P^\downarrow}{P^\uparrow} < 1$$

is satisfied as long as $H > (L - D)(1 + (\ln N - 1)^{-1/2})$. For example, for $N = 10^{10}$, the threshold condition becomes $H > 1.2(L - D)$. In other words, if the treatment-induced death rate exceeds the net growth rate in the absence of treatment by about 20 %, the pre-treatment phase plays the dominant role in the mutant generation.

The reason lies in the dynamics of the intermediate mutants. During the growth phase, a typical cell with a single mutation undergoes clonal expansion and this facilitates the generation of further mutations. During the treatment phase, a cell with a single mutation has a negative growth rate (as it is susceptible to one or more drugs) and this makes it unlikely that additional mutations can be attained before the clone is extinct.

We conclude the following:

- In the case of two drugs, the generation of resistant (double-hit) mutants in the pre-treatment phase strongly depends on the turnover rate, D/L. For increasing D/L, the probability of pre-existence is larger.
- We have $P^\uparrow > P^\downarrow$ even in the case where $H = H_c$. Therefore, in the case of two drugs, pre-existence again plays a more important role than the generation of resistant mutants during therapy. This effect is stronger than in the one-drug case. In principle, for any $H = (L - D)(1 + \eta)$, with $\eta > 0$, the (susceptible) tumor will eventually be eradicated. In the case of one drug, the pre-treatment phase is more important than the treatment phase only if $\eta > 1$. For two drugs, pre-treatment is more important as long as $\eta > (\ln(N/M_0) - 1)^{-1/2}$.

5.5 Summary: Evolutionary Dynamics of Resistance

All of the above arguments (and calculations) can be summarized as follows. For the case of **one drug**, the probability of treatment failure for a given size is largely independent of whether the cancer has a high or a low turnover rate. The contribution of the pre-treatment phase to the generation of resistance is greater than that of the treatment phase as long as the drug-induced death rate is higher than a threshold, $H > 2(L - D)$. On the other hand, for **two or more drugs**:

- Pre-treatment phase plays the dominant role in treatment failure (under some mild conditions on the drug-induced death rate, H, see above), and generation of resistance during treatment can be ignored, $P^\uparrow > P^\downarrow$ for the number of drugs in combination, $m > 1$;
- High-turnover cancers have a higher probability of treatment failure (for the same size N) than low-turnover cancers;
- Both of these effects become stronger for larger numbers of drugs.

5.6 Prevention of Resistance

After examining the basic evolutionary dynamics of drug resistance in cancer, we turn to a more applied question: How many drugs should be used to prevent treatment failure depending on the size of the tumor? We address the problem of treatment failure in the following way. We determine tumor size, N, at which the probability of treatment failure reaches a threshold value, denoted by δ. This means that if we start treatment at tumor size N, failure will be observed in a fraction δ of the patients, while treatment will be successful in a fraction $1 - \delta$ of patients. For example, let us assume that an acceptable goal is to treat 99 % of patients successfully, that is $\delta = 0.01$. Table 5.1 shows the tumor sizes at which resistance becomes a problem

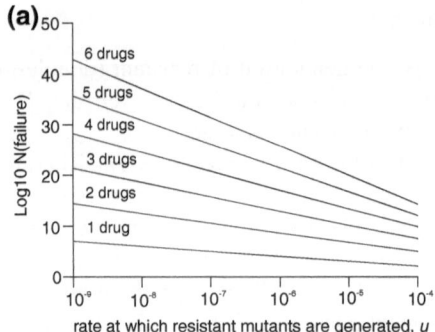

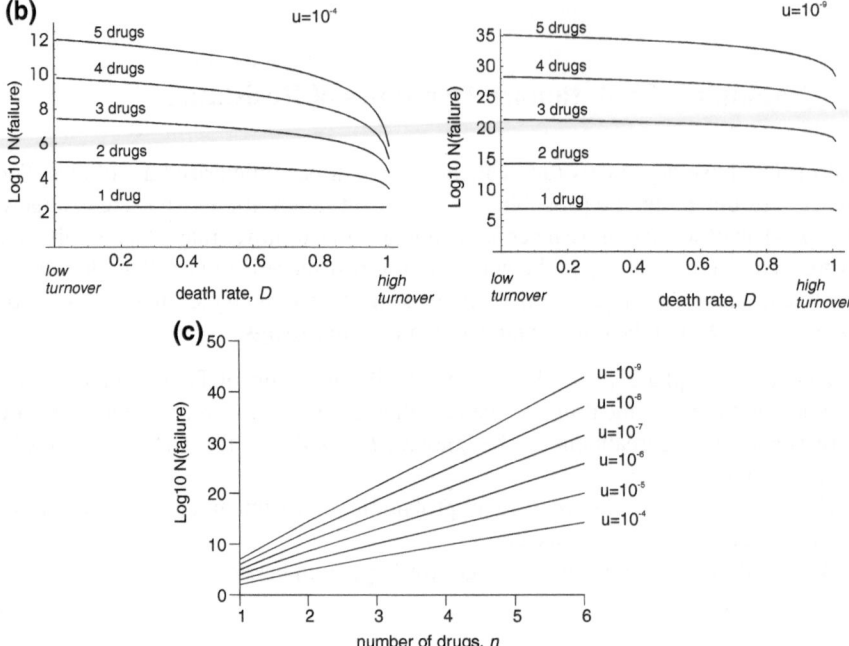

Fig. 5.3 Log tumor size, N, at which treatment failure is observed, depending on the parameters of the model. (**a**) Dependence on the rate at which resistant mutants are generated, u. The higher the value of u, the lower the tumor size at which treatment fails. The larger the number of drugs, the stronger this dependency. (**b**) Dependence on the natural death rate of tumor cells, D. The higher the value of D (i.e. the higher the turnover of the cancer), the lower the tumor size at which treatment fails. The higher the number of drugs, and the higher the rate at which resistant mutants are generated, u, the more pronounced this trend. (**c**) Dependence on the number of drugs, n. Increasing the number of drugs increases the tumor size at which treatment fails. The higher the mutation rate, however, the lower the advantage gained from adding further drugs. Baseline parameter values were chosen as follows: $L = 1$, $\delta = 0.01$

Table 5.1 The log10 size at which resistance becomes a problem (i.e. treatment fails in more than 1 % of patients), depending on the number of drugs and the rate at which resistant mutants are generated, u. If we assume that the cancers cannot grow beyond 10^{13} cells without causing death, a treatment regime can be considered acceptable if resistance only becomes a problem at sizes which are greater than 10^{13} cells (i.e. $\log_{10}$ of the size > 13). The parameter regimes where this occurs and treatment is expected to be successful are indicated by bold font in the table. The calculations assume $L = 1$, $D = 0$

	1 drug	2 drugs	3 drugs	4 drugs	5 drugs	6 drugs
$u = 10^{-4}$	2.01	4.95	7.46	9.81	12.06	**14.23**
$u = 10^{-5}$	3.01	6.73	10.13	**13.36**	**16.70**	**20.02**
$u = 10^{-6}$	4.01	8.61	12.91	**17.04**	**21.49**	**25.83**
$u = 10^{-7}$	5.01	10.53	**15.75**	**20.8**	**26.17**	**31.43**
$u = 10^{-8}$	6.01	12.47	**18.62**	**24.6**	**30.90**	**37.10**
$u = 10^{-9}$	7.01	**14.42**	**21.36**	**28.23**	**35.61**	**42.86**

(i.e. fewer than 99 % of patients will be treated successfully) depending on the rate at which resistance mutations are generated, u, and the number of drugs, m.

Assume that a single drug is used to treat patients. The tumor size when resistance arises is given by

$$\ln N = \ln \left[\frac{\delta(H + D - L)}{H} \right] - \ln u.$$

Resistance arises at lower tumor sizes for higher mutation rates, u (Fig. 5.3a). Note that if $H \gg D$, we have a very simple relation, $\ln N = \ln \delta - \ln u$. That is, the results are not influenced by the natural death rate, D. This is in accordance with our theoretical reasoning above.

If the number of drugs is increased, we observe three important differences:

1. An increase in the mutation rate, u, results in a more pronounced decline of the tumor size when resistance is observed. The larger the number of drugs, the more pronounced this decline (Fig. 5.3a).
2. The treatment phase, and thus the treatment efficacy, H, has no influence on the generation of resistance.
3. The size at which resistance arises now depends on the death rate, D. Resistance arises at smaller tumor sizes if the death rate of tumor cells, D, is higher (Fig. 5.3b). The larger the number of drugs and the higher the mutation rate, u, the stronger this dependency (Fig. 5.3b).

By how much does an increase in the number of drugs improve the chances of treatment success? According to the arguments above, this depends on the mutation rate, u, and the death rate of tumor cells, D.

(i) The higher the rate at which resistance mutations are acquired, u, the less is the effect of adding another drug, and the more difficult it becomes to treat (Fig. 5.3c). Consider the most optimistic scenario where $D = 0$ (Table 5.1). Assuming that cancers can reach up to sizes of 10^{13} cells [15], $u = 10^{-9}$ requires two drugs,

$u = 10^{-8} - 10^{-7}$ requires three drugs, $u = 10^{-6} - 10^{-5}$ requires four drugs, and $u = 10^{-4}$ requires six drugs (Table 5.1). By extrapolation, 10 drugs are needed if $u = 10^{-3}$, and about 30 drugs if $u = 10^{-2}$. Therefore, drugs to which resistance can be generated with such high rates (e.g. because genetic instability happens to promote the necessary mutations) should not be developed.

(ii) As pointed out above, resistance arises at lower tumor sizes as the death rate, D, is increased. In fact, if the death rate of tumor cells, D, comes close to their division rate, L (high-turnover cancer), then the effect of combining multiple drugs disappears (Fig. 5.3b). The size at which resistance arises converges to the same value, no matter how many drugs are used. In this case, the frequency with which cancers arise is low because they have a high chance to go extinct spontaneously, but when they do arise, the chances of complete tumor eradication are very slim. Because high-turnover cancers are likely to grow relatively slowly, however, drug therapy could still increase the life-span of the patient significantly by reducing the number of cells below a threshold rather than achieving a full response. Re-growth of resistant cells to large sizes would take a long time.

5.7 Application: CML

As a specific example, consider the treatment of CML [16] which is summarized in Chap. 2 in detail. Data suggest that two main types of mutations confer resistance to the cells [17–19]: the amplification of $BCR - ABL$, or a point mutation in the target protein. Genetic instability [20] is likely to promote the occurrence of gene amplifications which have been measured to occur in cancer cells at a rate of 10^{-4} per cell division [21]. On the other hand, the point mutation rate is about 10^{-9} per base per cell division [22]. Despite this fact, the frequency of gene amplifications is much less than that of point mutations among patients [18]. Part of the reason might be that $BCR - ABL$ amplifications are costly to the cells in the absence of treatment [23]. Including this assumption into the modeling framework, however, shows that even if this fitness cost is very significant, amplifications should still be observed more often than point mutations. Another consideration is that the level of resistance is a function of the number of extra copies of the $BCR - ABL$ gene. Therefore, if a significant degree of resistance requires two or more amplification events (but only one point mutation event), we expect that a resistant mutant is generated faster by point mutation than by gene amplification, explaining the observed frequencies.

Thus, for prevention of drug resistance we assume that resistant mutants are generated maximally with a point mutation rate of $u = 10^{-8} - 10^{-9}$. Experiments with susceptible CML cell lines have shown viability measurements (in the absence of treatment) of about 90 % [23]. From this we can roughly calculate that the relative death rate of cancer cells is in the range of $D/L = 0.1 - 0.5$. In addition, we also present calculations for $D/L = 0.9$. In Table 5.2 we show $\log_{10}$ size at which resistance becomes a problem, depending on the number of drugs and the turnover rate of the cancer cells (value of D/L). We consider treatment robust if resistance

Table 5.2 $\log_{10}$ size at which resistance becomes a problem, depending on the number of drugs and the turnover rate of the cancer cells (value of D/L)

	1 drug	2 drugs	3 drugs	4 drugs	5 drugs
$D/L = 0.1$	5.95	12.34	18.45	24.38	30.19
$D/L = 0.5$	5.95	12.13	17.99	23.69	29.26
$D/L = 0.9$	5.95	11.48	16.70	21.74	26.66

Table 5.3 Same as in Table 5.2, except that we assume that resistant mutants are generated with an elevated rate of $u = 10^{-6}$

	1 drug	2 drugs	3 drugs	4 drugs	5 drugs
$D/L = 0.1$	4.00	8.55	12.80	16.89	20.86
$D/L = 0.5$	4.00	8.31	12.37	16.20	19.93
$D/L = 0.9$	4.00	7.68	11.07	14.40	17.40

only arises at tumor sizes which are larger than 10^{13} cells (i.e. the value 13 in the table). In the relevant parameter region, we find that a combination of three drugs should prevent resistance and ensure successful therapy even for advanced cancers (Table 5.2). This assumes that the size of advanced cancers is less than 10^{13} cells, which derives from white blood cell count measurements which range from $10^5 - 10^6$ per microliter of blood in blast crisis.

In Table 5.2 we assume that resistant mutants are generated with a rate of $u = 10^{-8}$. The reason for this parameter choice is as follows: while the point mutation rate is around $u = 10^{-9}$, several point mutations can lead to resistance and this increases the rate. The findings of [24] indicate that $BCR - ABL$ might increase the amount of reactive oxygen species and thus the rate of point mutations. In Table 5.3 we consider an 100-fold elevated mutation rate $u = 10^{-6}$. This represents the borderline where three drugs will not prevent resistance anymore. We conclude that as long as the elevation of the mutation rate is less than a 100-fold, our results remain robust.

Note that our calculations assumed that drug-resistant cells have the same fitness as susceptible cells in the absence of treatment. Thus, our estimates are conservative, and the number of required drugs might be lower if drug-resistant mutants are characterized by a certain fitness cost.

5.8 Model Extension and Applicability

In addition to imatinib, possible candidates for additional drugs to be used in combination in CML therapy have been discussed in the literature [25], although the most promising ones show some degree of cross-resistance with imatinib [26]. If this is the case, our framework still applies, but the calculations have to be modified as described extensively in Chap. 8.

Another important issue is the heterogeneity of tumors. In CML (as well as AML, and several solid tumors including breast and central nervous system tumors [27–29]), there is evidence for the existence of cancer stem cells, comprising a fraction of the total tumor burden. For CML, the fraction of stem cells in blast crisis is more that 30 %, and it is much smaller in the chronic phase [27–29]. It has been proposed that these cancer stem cells, which are the only tumor cells that have potential for self-renewal, may account for drug resistance after initial response to therapy. This circumstance can be taken into account by using the present framework. As resistance is mainly a problem in blast crisis and usually does not arise in the chronic phase, we performed our calculations for the latter phase of the disease. During this phase, the blasts undergo a phase of rapid exponential growth, and therefore the quantitative results of our present calculations apply. However, it is important to consider heterogeneous populations of the chronic and accelerated phases of CML and to model stem cell behavior, which is the subject of Chap. 6.

In the present calculations we assumed that resistant mutants behave in the same way as the wild type tumor cells before treatment starts. This may not be the case. If one can establish that resistant mutants possess a fitness advantage in the absence of treatment, this will definitely make the estimate of the probability of resistance generation higher. Indeed, resistant mutants will grow faster and reach higher numbers (and a larger fraction of the total tumor load) before the treatment starts. On the other hand, if resistant mutants are at a disadvantage before the beginning of therapy, this would make generation of resistance less likely.

5.9 Summary

This chapter has shown how a mathematical framework can be used to understand the principles that underlie the emergence of resistance in cancers treated with targeted drugs, such as in the treatment of CML with imatinib and other small molecule inhibitors. The model clearly showed that resistance is generated during the pretreatment growth phase of the cancer, and that the treatment phase is irrelevant for the generation of resistant mutants. While in the context of treatment with a single drug, the turnover of the tumor cell population did not influence the evolution of drug-resistant mutants, the turnover did have a significant effect in the context of treatment with two or more targeted drugs. Hence, it would be important to measure the cellular turnover kinetics at different stages of the disease. The rate of cell division, L, and the rate of cell death, D, can be calculated from DNA labeling data, similar to studies performed in the context of immunology [30]. This framework was applied to study how a combination of different small molecule inhibitors can prevent treatment failure induced by drug resistance. In the context of CML, available parameter estimates suggest that a combination of three drugs could ensure successful therapy, even at advanced stages of the disease. Note that this is a conservative estimate and that the number of required drugs might be lower if resistant mutants carry a fitness cost. The following chapters will elaborate on the work described here and incorporate further biological complexities into the model to examine how this affects the evolution of drug resistant mutants, and how this can be clinically managed.

Problems

5.1. Derive exact conditions under which the approximation by an exponent in formula (5.1) is valid.

5.2. Research Project
Find out about the history of small molecule inhibitors, the discovery of imatinib, and the next generation CML drugs such as dasatinib and nilotinib.

5.3. Research Project
Find out which small molecule inhibitors are currently available in the market for the treatment of different cancers (see also Sect. 2.1). Which cancers can be thought of being low-turnover and high-turnover cancers? What consequences may this have for the treatment success of these cancers?

5.4. Research Project
What are the similarities and differences with resect to the emergence of drug resistance in cancer and in HIV therapies? Find out about combination treatments for HIV.

5.5. Research Project
How would the modeling approach be modified if we wanted to describe the generation of drug resistance by gene amplification, rather than by point mutations?

5.6. Numerical Project
Use the program obtained for numerical project 4.8 to investigate how the probability of treatment success depends on the turnover D/L, for different numbers of drugs m. What do you observe as D/L becomes close to 1? How do properties of the system change as m increases?

References

1. Fojo, T., Bates, S.: Strategies for reversing drug resistance. Oncogene **22**(47), 7512–23 (2003)
2. Gottesman, M.: Mechanisms of cancer drug resistance. Annu. Rev. Med. **53**(1), 615–627 (2002)
3. Luqmani, Y.: Mechanisms of drug resistance in cancer chemotherapy. Med. Princ. Pract. **14**(1), 35–48 (2005)
4. Redmond, K., Wilson, T., Johnston, P., Longley, D.: Resistance mechanisms to cancer chemotherapy. Front Biosci. **13**, 5138–5154 (2008)
5. Ozben, T.: Mechanisms and strategies to overcome multiple drug resistance in cancer. FEBS Lett. **580**(12), 2903–2909 (2006)
6. Longley, D., Johnston, P.: Molecular mechanisms of drug resistance. J. Pathol. **205**(2), 275–292 (2005)
7. Druker, B.J.: Imatinib as a paradigm of targeted therapies. Adv. Cancer Res. **91**, 1–30 (2004)

8. Blagosklonny, M.V.: Sti-571 must select for drug-resistant cells but 'no cell breathes fire out of its nostrils like a dragon'. Leukemia **16**(4), 570–572 (2002)
9. Shannon, K.M.: Resistance in the land of molecular cancer therapeutics. Cancer Cell **2**(2), 99–102 (2002)
10. Khorashad, J., Kelley, T., Szankasi, P., Mason, C., Soverini, S., Adrian, L., Eide, C., Zabriskie, M., Lange, T., Estrada, J., et al.: Bcr-abl1 compound mutations in tyrosine kinase inhibitor resistant cml: frequency and clonal relationships. J. Am. Soc. Hematol. **119**(10), 2234–2238 (2012)
11. Simon, V., Ho, D.D.: Hiv-1 dynamics in vivo: implications for therapy. Nat. Rev. Microbiol. **1**(3), 181–190 (2003)
12. Schabel, F.M. Jr., Skipper, H.E., Trader, M.W., Laster, W.R. Jr., Griswold, D.P. Jr., Corbett, T.H.: Establishment of cross-resistance profiles for new agents. Cancer Treat. Rep. **67**(10), 905–922 (1983)
13. Schabel, F.M. Jr., Trader, M.W., Laster, W.R. Jr., Corbett, T.H., Griswold, D.P. Jr.: cis-dichlorodiammineplatinum(ii): combination chemotherapy and cross-resistance studies with tumors of mice. Cancer Treat. Rep. **63**(9–10), 1459–1473 (1979)
14. Coldman, A.J., Goldie, J.H.: A stochastic model for the origin and treatment of tumors containing drug-resistant cells. Bull. Math. Biol. **48**(3–4), 279–292 (1986)
15. McKinnell, R.G., Parchment, R.E., Perantoni, A.O., Pierce, G.B.: The Biological Basis of Cancer. Cambridge University Press, Cambridge (1998)
16. Calabretta, B., Perrotti, D.: The biology of cml blast crisis. Blood **103**(11), 4010–4022 (2004)
17. McCormick, F.: New-age drug meets resistance. Nature **412**(6844), 281–282 (2001)
18. Gambacorti-Passerini, C.B., Gunby, R.H., Piazza, R., Galietta, A., Rostagno, R., Scapozza, L.: Molecular mechanisms of resistance to imatinib in philadelphia-chromosome-positive leukaemias. Lancet Oncol. **4**(2), 75–85 (2003)
19. Gorre, M.E., Mohammed, M., Ellwood, K., Hsu, N., Paquette, R., Rao, P.N., Sawyers, C.L.: Clinical resistance to sti-571 cancer therapy caused by bcr-abl gene mutation or amplification. Science **293**(5531), 876–880 (2001)
20. Loeb, L.A.: Cancer cells exhibit a mutator phenotype. Adv. Cancer Res. **72**, 25–56 (1998)
21. Tlsty, T.D., Margolin, B.H., Lum, K.: Differences in the rates of gene amplification in nontumorigenic and tumorigenic cell lines as measured by luria-delbruck fluctuation analysis. Proc. Natl. Acad. Sci. U.S.A. **86**(23), 9441–9445 (1989)
22. Loeb, L.A., Springgate, C.F., Battula, N.: Errors in dna replication as a basis of malignant changes. Cancer Res. **34**(9), 2311–2321 (1974)
23. Tipping, A.J., Mahon, F.X., Lagarde, V., Goldman, J.M., Melo, J.V.: Restoration of sensitivity to sti571 in sti571-resistant chronic myeloid leukemia cells. Blood **98**(13), 3864–3867 (2001)
24. Nowicki, M.O., Falinski, R., Koptyra, M., Slupianek, A., Stoklosa, T., Gloc, E., Nieborowska-Skorska, M., Blasiak, J., Skorski, T.: Bcr/abl oncogenic kinase promotes unfaithful repair of the reactive oxygen species-dependent dna double-strand breaks. Blood **104**(12), 3746–3753 (2004)
25. Yoshida, C., Melo, J.V.: Biology of chronic myeloid leukemia and possible therapeutic approaches to imatinib-resistant disease. Int. J. Hematol. **79**(5), 420–433 (2004)
26. Shah, N.P., Tran, C., Lee, F.Y., Chen, P., Norris, D., Sawyers, C.L.: Overriding imatinib resistance with a novel abl kinase inhibitor. Science **305**(5682), 399–401 (2004)
27. Faderl, S., Talpaz, M., Estrov, Z., Kantarjian, H.M.: Chronic myelogenous leukemia: biology and therapy. Ann. Intern. Med. **131**(3), 207–219 (1999)
28. O'Dwyer, M.E., Mauro, M.J., Druker, B.J.: Recent advancements in the treatment of chronic myelogenous leukemia. Annu. Rev. Med. **53**, 369–381 (2002)
29. Al-Hajj, M., Clarke, M.F.: Self-renewal and solid tumor stem cells. Oncogene **23**(43), 7274–7282 (2004)
30. Asquith, B., Debacq, C., Macallan, D.C., Willems, L., Bangham, C.R.: Lymphocyte kinetics: the interpretation of labelling data. Trends Immunol. **23**(12), 596–601 (2002)

Chapter 6
Effect of Cellular Quiescence on the Evolution of Drug Resistance in CML

Abstract This chapter examines the role of tumor stem cell dynamics on the evolution of drug resistant mutants in the context of CML therapy with small molecule inhibitors. In particular, the role of tumor stem cell quiescence is examined. Modeling shows that in the context of treatment with a single drug, parameters that determine the kinetics of cellular quiescence do not affect the probability of treatment failure as a result of drug resistant mutants. On the other hand, if two or more drugs are used in combination to treat the cancer, then treatment failure as a result of drug resistance is promoted by the occurrence of cellular quiescence. Interestingly, while cellular quiescence significantly prolongs the time until the cancer has dropped to low numbers or has been driven extinct, the model predicts that drug resistance does not evolve during this treatment phase. The occurrence of cellular quiescence increases the likelihood that resistant mutants are generated during the growth phase of the cancer before therapy is initiated.

Keywords Cancer treatment with small molecule inhibitors · Combination therapies · Multiple-drug therapies · Cellular quiescence · Stem cells · Cycling cells · Stochastic modeling · Deterministic modeling · Evolution of resistance · Pre-treatment phase · Probability of treatment failure · Pre-existence of resistant mutants

6.1 Introduction

The previous chapter examined the evolution of CML cell variants that are resistant to targeted drugs, such as imatinib. This was done in the context of the simplest model that assumed exponential growth of the tumor cells and a simple birth-death process, where cell division and cell death are the major determinants of the dynamics. Chapter 3, however, has shown that the dynamics of CML growth can be more complex and that this can influence the treatment dynamics. In particular, stem cell dynamics and the distinction between active and quiescent tumor stem cells could be important key factors influencing the dynamics. The same is

N. L. Komarova and D. Wodarz, *Targeted Cancer Treatment in Silico*,
Modeling and Simulation in Science, Engineering and Technology,
DOI: 10.1007/978-1-4614-8301-4_6, © Springer Science+Business Media New York 2014

likely to apply to the evolutionary dynamics of drug resistant cell variants. This chapter explores how the ability of tumor stem cells to enter quiescent and activated states can affect the generation of mutants that are resistant to targeted drugs. Results are compared to those obtained for the simpler model in the previous chapter.

6.2 Modeling Resistance Generation in the Context of Quiescence

We formulate a stochastic model that includes a population of primitive, proliferating CML cells, and a population of quiescent CML cells. The proliferating cells divide with a rate L and die with a rate d. The death rate captures both the natural death rate of cancer cells and the treatment-induced death rate. In the absence of treatment, $L > d$ and the cell population grows exponentially. Treatment increases the parameter d. If treatment is efficient, then $L < d$, such that the tumor cell population declines. The cells enter a quiescent state with a rate α, and quiescent cells re-enter the cell cycle with a rate β. Note that quiescent cells do not divide or die and are not susceptible to any drug activity. In addition to these processes, CML cells can mutate to give rise to acquired drug resistance. In particular, we assume that during cell division, a resistant mutant is generated with a probability u. We further assume that CML cells grow exponentially to a defined size N, after which the disease is detected and imatinib therapy is started. We calculate the probability that the cancer is driven extinct by therapy, i.e., the probability that no resistant mutants spread before the CML cells have gone extinct. We examine how this probability depends on the parameters that determine cellular growth, mutations, quiescence, and death. When talking about tumor extinction in the model, we always imply extinction brought about by drug therapy. This should be thought of as "clinical remission" in medical rather than mathematical terms.

6.3 Methodology

Here we use the methodology developed in Chap. 4 and include two more processes to our description of cell colony dynamics: the process of becoming a quiescent cell (rate α), and a process of returning from the state of quiescence (rate β).

6.3.1 The Basic Formalism of Quiescence

Let us denote by i the number of cycling cells and by j the number of quiescent cells. In the absence of mutations, the system's state is defined by the pair (i, j). The dynamics of cells can be described by a following continuous time birth-death process. In an infinitesimally small interval of time, Δt, only the following changes in the system are possible:

- $(i, j) \rightarrow (i + 1, j)$ with probability $iL\Delta t$, a cell division;
- $(i, j) \rightarrow (i - 1, j)$ with probability $id\Delta t$, a cell death;
- $(i, j) \rightarrow (i - 1, j + 1)$ with probability $i\alpha\Delta t$, a cell turning quiescent;
- $(i, j) \rightarrow (i + 1, j - 1)$ with probability $j\beta\Delta t$, a quiescent cell waking up;
- $(i, j) \rightarrow (i, j)$ with probability $1 - (i[L + d + \alpha] + j\beta)\Delta t$, no change.

The Kolmogorov forward equation for this system is:

$$\dot{\varphi}_{ij} = L(i - 1)\varphi_{i-1} + d(i + 1)\varphi_{i+1} + \alpha(i + 1)\varphi_{i+1,j-1} + \beta(j + 1)\varphi_{i-1,j+1}$$
$$- \varphi_{ij}[(L + d + \alpha)i + \beta j]. \tag{6.1}$$

Suppose that initially, the number of cycling cells is given by I_0 and the number of quiescent cells is given by J_0, i.e. $\varphi_{I_0, J_0}(0) = 1$, and $\varphi_{ij}(0) = 0$ for all other values of i and j.

Let us denote by x the average number of cycling cells, and y the average number of quiescent cells:

$$x = \sum_{i,j=1}^{\infty} \varphi_{ij} i, \quad y = \sum_{i,j=1}^{\infty} \varphi_{ij} j. \tag{6.2}$$

The equations for x and y are as follows:

$$\dot{x} = (L - d - \alpha)x + \beta y, \tag{6.3}$$
$$\dot{y} = \alpha x - \beta y, \tag{6.4}$$
$$x(0) = I_0, \quad y(0) = J_0. \tag{6.5}$$

Please note that these equations can be used, for example, to study the phenomenon of the biphasic decline, Chap. 3.

6.3.2 Modeling Resistance to Drugs in a Population with Quiescence: A Stochastic Approach

A cell can be either resistant or susceptible to each of the m drugs. In total, there are $2^m - 1$ resistant types. As before, they can be enumerated by a binary index s, which is a string of m components with numerical values from the set $\{0, 1\}$.

We denote by i_s the number of cycling cells of resistance type s, and j_s the number of quiescent cells of resistance type s. Each type s is characterized by its division rate, L_s, and death rate, d_s. Again, all the types are naturally placed on a binary directed network where each edge connects two types which only differ by their resistance properties with respect to one drug. A single mutation separates each pair of connected types. The Kolmogorov forward equation for this system can be written (not shown). The following probability generating function can be defined:

$$\Psi(\xi_0, \eta_0, \ldots, \xi_n, \eta_n; t) = \sum_{s=0}^{n} \varphi_{i0, j0, \ldots, in, jn} \xi_0^{i_0} \eta_0^{j_0} \ldots \xi_n^{i_n} \eta_n^{j_n}.$$

The probability generating function satisfies the following PDE:

$$\frac{\partial \Psi}{\partial t} = \sum_{s=0}^{n} \left(\frac{\partial \Psi}{\partial \xi_s} \left(L_s (1 - \sum_j u_{s \to j}^{out}) \xi_s^2 + \sum_j L_s u_{s \to j}^{out} \xi_j^{out} + d_s + \alpha_s \eta_s \right. \right.$$

$$\left. \left. -(L_s + d_s + \alpha_s) \xi_s \right) + \frac{\partial \Psi}{\partial \eta_s} \beta_s (\xi_s - \eta_s) \right).$$

For each type s, the quantities $u_{s \to j}$ are the mutation rates corresponding to the mutations of type s into each of the types j, with which s is directly connected. The types j are assumed to be downstream from type s. The linear transport-type PDE can be solved by the standard method of characteristics:

$$\dot{\xi}_s = L_s (1 - \sum_j u_{s \to j}^{out}) \xi_s^2 + \sum_j L_s u_{s \to j}^{out} \xi_j^{out} + d_s + \alpha_s \eta_s - (L_s + d_s + \alpha_s) \xi_s,$$

$$(6.6)$$

$$\dot{\eta}_s = \beta_s (\xi_s - \eta_s). \tag{6.7}$$

Using symmetry assumptions similar to those in Eq. (4.11), we can write down the following equations for characteristics:

$$\dot{\xi}_k = L_k (1 - (m - k)u) \xi_k^2 + [(m - k)L_k u \xi_{k+1} - (L_s + d_k + \alpha_k)] \xi_k + d_k + \alpha_k \eta_k,$$

$$(6.8)$$

$$\dot{\eta}_k = \beta_k (\xi_k - \eta_k), \quad 0 \le k \le m. \tag{6.9}$$

The coefficients in Eqs. (6.8–6.9) depend on whether treatment takes place or not. We assume that before the start of treatment, the growth and death rates of all types are the same:

$$L_k = L, \quad d_k = D < L, \quad 0 \le k \le m.$$

As before, we assume that after the treatment starts, the fully susceptible and all the partially resistant types are killed by the drugs with the intensity H, and the fully resistant type is not affected by treatment. This translates into the following values for the coefficients:

$$L_k = L \; \forall k, \quad d_k = D + H > L, \quad 0 \le k \le m - 1, \quad d_m = D.$$

Let us suppose that at the start of treatment we have I_0 cells of class $0, \ldots, I_m$ cells of class m. Then the probability of eventual colony extinction (related to the long-term treatment success by equation of type (4.10)) is given by

$$P_{\text{succ}} = \lim_{t \to \infty} \Psi(0, \ldots, 0; t) = \prod_{k=0}^{m} [\xi_k^{\infty}]^{I_k}, \qquad (6.10)$$

where $\xi_k^{\infty} = \lim_{t \to \infty} \xi_k(t)$. From studying the fixed points of system (6.8–6.9), we can see that

$$\xi_k^{\infty} = \eta_k^{\infty},$$

and the Eq. (6.8) at steady state do not depend on quiescence parameters. Therefore the limiting values ξ_k^{∞} do not depend on the quiescence parameters α and β. This means that the probability of treatment success (for a given colony) is not affected by the process of quiescence in the treatment regime (but it strongly depends on the number distribution of mutations at the start of treatment, Eq. (6.10)).

To relate the colony's age (t_*) and size (N) at the start of treatment, similar to Eq. (4.9), we set $x(t_*) + y(t_*) = N$, where the functions $x(t)$ and $y(t)$ satisfy the system (6.3–6.4) with the initial condition $x(0) = M_0$, $y(0) = 0$.

6.3.3 Resistance Generation: A Deterministic Approach

Another way to study the production of resistant mutants is to look at the expected behavior (the first moments) of various types of cells. We can derive equations for the moments from the Kolmogorov forward equations.

Equations for the moments. To gain intuitive understanding of the dependence of mutant production upon quiescence, we consider a simplified system for m drugs with $d_s = 0$, and symmetric rates $L_s = L$, $\alpha_s = \alpha$, $\beta_s = \beta$:

$$\dot{x}_0 = [L(1 - u_0) - \alpha]x_0 + \beta y_0, \qquad (6.11)$$
$$\dot{y}_0 = \alpha x_0 - \beta y_0, \qquad (6.12)$$
$$\cdots \qquad (6.13)$$
$$\dot{x}_k = L u_{k-1} x_{k-1} + [L(1 - u_k) - \alpha]x_k + \beta y_k, \qquad (6.14)$$
$$\dot{y}_k = \alpha x_k - \beta y_k, \quad 0 < k \le m. \qquad (6.15)$$

Here, the rates u_k are probabilities of mutation from class k to class $k + 1$. Because of the combinatorial nature of our problem, we have

$$u_k = (m - k)u,$$

where u is a basic rate of mutations.

The expected number of mutants for $\beta = 0$. Let us assume that $\beta = 0$. Then linear system (6.11–6.15) can be solved analytically (the simplifying assumption that $\beta = 0$ makes the equations for cycling cells independent of the equations for quiescent cells). In particular, for any value of k, we have the exact solution

$$\sum_{k=0}^{m}(x_k + y_k) = \frac{Le^{(L-\alpha)t} - \alpha}{L - \alpha}.$$

Therefore the time it takes to reach size N is given by

$$t_* = \frac{\ln\left[\frac{N(L-\alpha)+\alpha}{L}\right]}{L - \alpha}. \tag{6.16}$$

To calculate the fraction of cycling cells of class k at size N, we substitute the above expression for t_* into the exact solution for $x_m(t) + y_m(t)$, and expand in a double Taylor series in terms of small parameters u and $1/N$. We obtain:

$$\frac{x_k}{x_k + y_k} = \left(1 - \frac{\alpha}{L}\right)(1 + O(1/\ln N)) + O(u). \tag{6.17}$$

Similarly, the average number of mutants resistant to m drugs is given by

$$x_m(t_*) + y_m(t_*) \approx N\left(\frac{u\ln N}{1 - \frac{\alpha}{L}}\right)^m. \tag{6.18}$$

Again, here we only present the first term in the u-expansion and the largest term in the $1/\ln N$ expansion of the exact expressions.[1] Exact values of quantities $x_k(t_*) + y_k(t_*)$ and $\frac{x_k}{x_k + y_k}$ are plotted in Fig. 6.1a, b as functions of α.

Let us first consider the case $m = 1$. The average number of (cycling and quiescent) 1-hit mutants in this case is an increasing function of α, see Eq. (6.18). On the other hand, the average number of *cycling* 1-hit mutants is obtained from Eqs. (6.18) and (6.17) and is given by $Nu\ln N$, that is, it is independent of α. To understand this result, we note that cycling mutants are produced by cycling wild-type cells and they grow according to the same law as the cells producing them. When α increases, the mutant clones grow more slowly because of quiescence, but at the same time they have more time to grow (it takes more "events" before the number of cells reaches size N). In other words, the changes in the mutant growth are completely compensated by the change in the time of growth. That is why the quantity $x_1(t_*)$ is a function of the colony size only, see Fig. 6.1(c).

The dependence of the total number of mutants, $x_m(t_*) + y_m(t_*)$, on α becomes stronger as the number of drugs increases, see Fig. 6.1. In particular, both the total

---

[1] In order to perform the Taylor series expansion we assume that $u \ll 1 - \alpha/L$; for α very close to L, the value $1 - \alpha/L$ would have to be taken to be the small parameter of expansion.

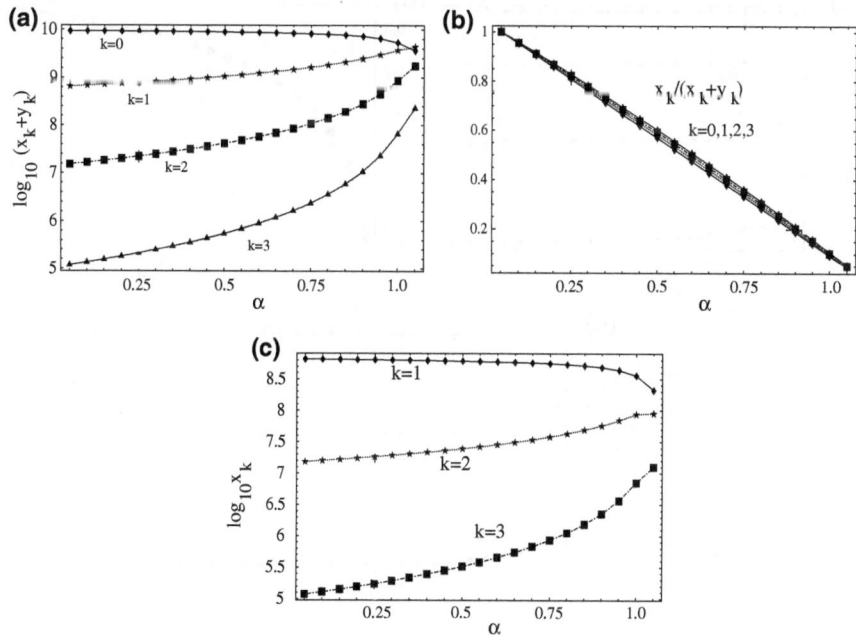

Fig. 6.1 A numerical illustration of the dependence of solutions of (6.11–6.15) on α, with $\beta = 0$. (a) The total number of mutants of class k, $x_k + y_k$; (b) The fraction of cycling mutants of class k, $\frac{x_k}{x_k + y_k}$; (c) The number of cycling mutants of class k, x_k. The parameters are $N = 10^{10}$, $m = 3$, $L = 1$, $u = 10^{-3}$

number of 2-hit mutants and the number of cycling 2-hit mutants depend on α, formula (6.18).

The probability of having resistance to m drugs, $\beta = 0$. In our studies of resistance generation, two quantities are of particular importance: the expected number of resistant mutants, and the probability of having resistance. Of course, the latter quantity (the probability of having resistance) cannot be obtained directly when studying system (6.11–6.15). However, we can study a related quantity, $W_{m-1} \equiv \int_0^{t_*} L u x_k(t)\, dt$, the total number of fully resistant mutants produced by the colony. We have

$$W_{m-1} = mNu \left(\frac{u \ln n}{1 - \frac{\alpha}{L}} \right)^{m-1}. \tag{6.19}$$

In particular, for $m = 1$, both W_0 and the probability of generating resistance to one drug is independent of α. However, the dependence on α does take place for larger values of m. For example, the probability of resistance to two drugs is an increasing function of α.

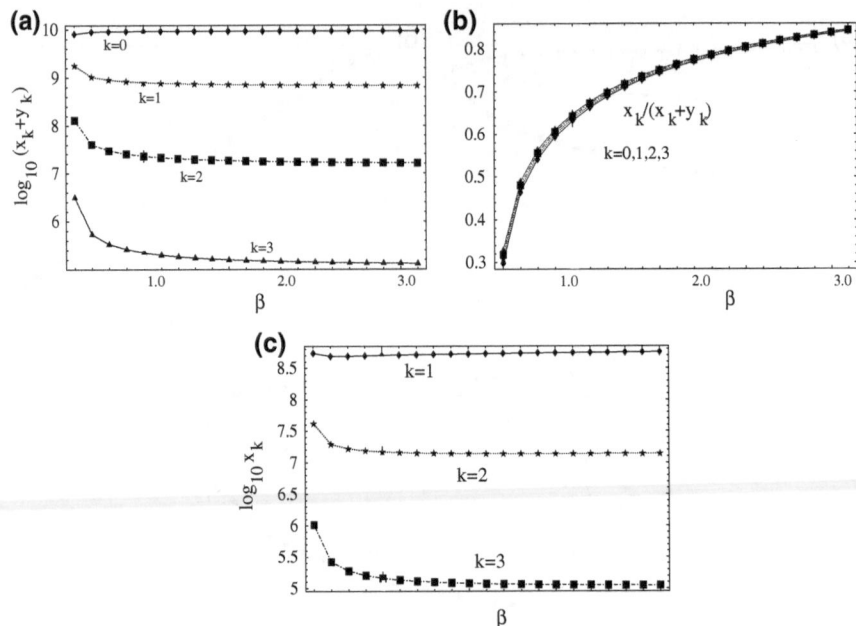

Fig. 6.2 The dependence of solutions of (6.11–6.15) on β. (**a**) The total number of mutants of class k, $x_k + y_k$; (**b**) The fraction of cycling mutants of class k, $\frac{x_k}{x_k + y_k}$; (**c**) The number of cycling mutants of class k, x_k. The parameters are as in Fig. 6.1, with $\alpha = 0.7$

The case of $m = 2$ drugs can be explained intuitively. Two-hit mutants are produced by cycling one-hit mutants. At each value of N, the number of cycling wild-type cells is given by

$$x_0(t_*) = (1 - \alpha/L)N(1 + o(u)),$$

that is, it decreases with α, and the number of cycling one-hit mutants is

$$x_1(t_*) = mNu \ln N(1 + o(u)),$$

which is α-independent. The last two expressions are obtained by solution system (6.11–6.15) directly and using Eq. (6.16). We can see that the relative proportion of one-hit mutants for a given colony size is a growing function of α. Therefore, for increasing α, every time a cell divides, it becomes more likely that the dividing cell already has a mutation. Therefore the production of two-hit mutants is a growing function of α, see the expression for W_1, Eq. (6.19).

The case $\beta > 0$. Once we include a nonzero value of β in system (6.11–6.15), the equation $\sum_{k=0}^{m}(x_k + y_k) = N$ cannot be analytically resolved for t_*. In other words, there is no equivalent of formula (6.16) with $\beta > 0$. In order to see how the results depend on the value of β, we have performed numerical solutions of system (6.11–6.15). These are presented in Fig. 6.2. We can see that the total number of mutants of class k decays with β (Fig. 6.2a), and the fraction of cycling cells increases with β (Fig. 6.2b). The number of cycling mutants, x_k, is a decaying function of β for $k > 1$ (Fig. 6.2c).

These results show that the dependence of the number of resistant mutants on the rate of cell awakening, β, is the opposite of its dependence on the rate of quiescence, α. In general, the number of resistant mutants increases with the amount of quiescence in the system. The probability of resistance to k drugs (which is related to the function x_{k-1}, formula (6.19)), also increases with the amount of quiescence, except for the case $k = 1$, where it is independent of the amount of quiescence.

6.4   Basic Evolutionary Dynamics: Growth Versus Treatment Phase

In the previous chapter that did not take into account cellular quiescence, the result was obtained that the treatment phase is largely irrelevant for the generation of resistance. That is, if treatment does fail because of drug resistant mutants, these mutants were likely generated in the growth phase before the start of therapy. Quiescence can significantly slow down the rate with which the tumor cell population declines during treatment, thus prolonging this phase. The argument has been made that the tumor might acquire resistance during this phase and that this could lead to a relapse of the tumor after a certain time, despite continued therapy. Performing appropriate analysis with the current model (see Sect. 6.3.2), it is found that even in the presence of quiescence, the treatment phase is not relevant for the generation of drug resistant mutants, no matter how long treatment takes. Thus, if at the start of therapy no resistant mutants exist, treatment is likely to result in the extinction of the tumor, given enough time. Therefore, strategies aimed at shortening the treatment phase, for example by activating quiescent cells, will not reduce the chances that treatment fails as a result of drug resistance.

6.5   Quiescence and the Evolution of Resistance

Keeping in mind that the tumor growth phase drives resistance evolution rather than the treatment phase, we calculate the probability of therapy success depending on the rate at which cells enter quiescence, α, and the rate at which cells exit the quiescent state, β. Several scenarios are considered. First, we study resistance against a single drug (i.e., imatinib in CML treatment). We then also take into account resistance against two or more drugs used in combination. Throughout the next few paragraphs, we make the simplifying assumption that the cell death rate in the pretreatment phase is zero. Also, the theoretical explanations will concentrate on one of the quiescence parameters, α, which is the rate of entering the state of quiescence. The rate of cell awakening, β, can be treated similarly. Figure 6.3 illustrates the α- and β- dependence of the probability of no resistance. It was created by numerical solutions of ordinary

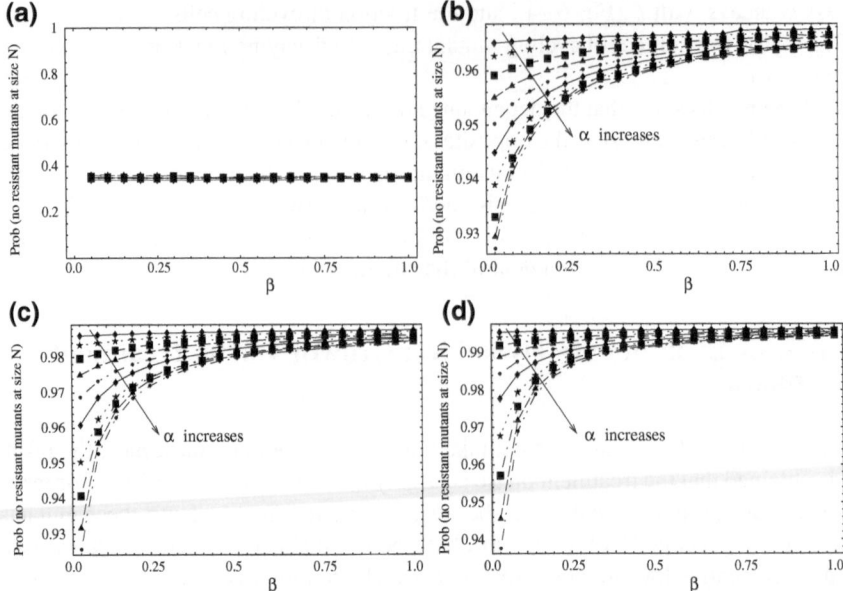

Fig. 6.3 The probability of having no fully resistant mutants at size N for different quiescence parameters. Each figure (**a–d**) shows the probability of no resistant mutants as a function of β (the rate of cell awakening), for 10 different values of α (the rate at which cells become quiescent), $\alpha = 0.1, 0.2, \ldots$ and 1.0. (**a**) Treatment with $m = 1$ drugs; all the curves corresponding to different values of α are the same. The parameters are $N = 10^7$ and $u = 10^{-7}$. (**b**) Treatment with $m = 2$ drugs, $N = 10^{11}$, $u = 10^{-7}$. (**c**) $m = 3$ drugs, $N = 10^{13}$, $u = 10^{-6}$. (**d**) $m = 4$ drugs, $N = 10^{13}$, $u = 10^{-5}$. In all plots, we took $M_0 = 10^3$, $L = 1$, $d = 0$. The reason we used different values of N and u for different values of m is because we chose the parameter regime corresponding to intermediate values of the probability of treatment success. When this probability is nearly 100 % or nearly 0, then the dependence on α and β is less apparent and less meaningful

differential equations for the characteristics, see the Chap. 4 for mathematical details. The calculations give rise to the following findings.

6.5.1 Probability of One-Drug Treatment Failure (Due to Resistance) is Independent of Quiescence

The probability to observe treatment failure as a result of resistance in the context of a single drug is not affected by quiescence parameters (Fig. 6.3a). To put this in quantitative terms, the probability to have at least one resistant mutant at size N is independent of α and β.

This is demonstrated by the following argument. Let us assume for simplicity that there is no cell death in the colony (all the arguments can be extended to nonzero

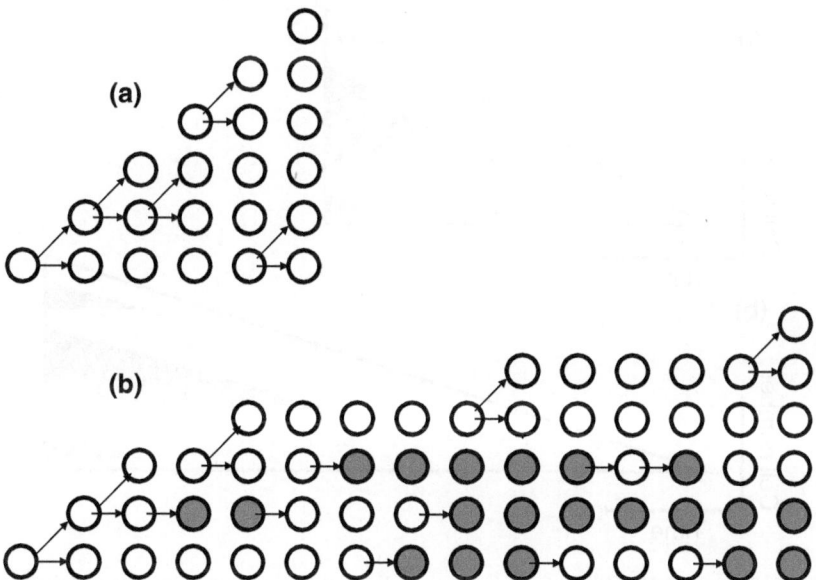

Fig. 6.4 A schematic demonstrating the number of cell divisions that is needed for a colony of cells to expand from 1 cell to N cells (in the figure, $N = 6$). Empty circles represent cycling cells, and gray circles represent quiescent cells. Columns depict states of the colony in consecutive moments of time. The changes are marked by arrows. Two arrows stemming from one cell represent a cell division. A single arrow represents either a cell becoming quiescent or a quiescent cell waking up. (a) A colony without quiescence. (b) A colony with quiescence. In both cases, we can see that it takes exactly $N - 1 = 5$ cell divisions to expand to size N; however, the process in (b) contains more "events" and it takes a longer time

death rates). In the model, mutants are generated during cell division. The probability of resistance is the same as the probability to generate mutants, which is defined by the number of cell divisions (and the constant mutation rate). It is easy to see that the total number of cell divisions until the tumor reaches size N does not depend on the quiescence parameters α and β. For instance, if there is no cell death, then the number of cell divisions to expand from one cell to N cells is exactly N-1, no matter what the quiescence rates are, see Fig. 6.4. It is of course the case that the higher the rate at which cells enter quiescence, and the lower the rate at which cells exit quiescence, the longer it takes the tumor to grow to size N. However, the actual number of cell divisions to reach size N is unchanged by quiescence. Therefore the probability to produce resistant mutants is independent of quiescence rates.

As we will see in the following paragraphs, the situation is different when considering resistance against two or more drugs. For treatment with multiple drugs, the probability of treatment failure as a result of resistance depends on the quiescence parameters (Fig. 6.3b–d). The higher the rate of entry into the quiescent state (larger α) and the lower the rate of exit from the quiescent state (lower β), the higher the probability of treatment failure. In order to explain this, we will consider generating

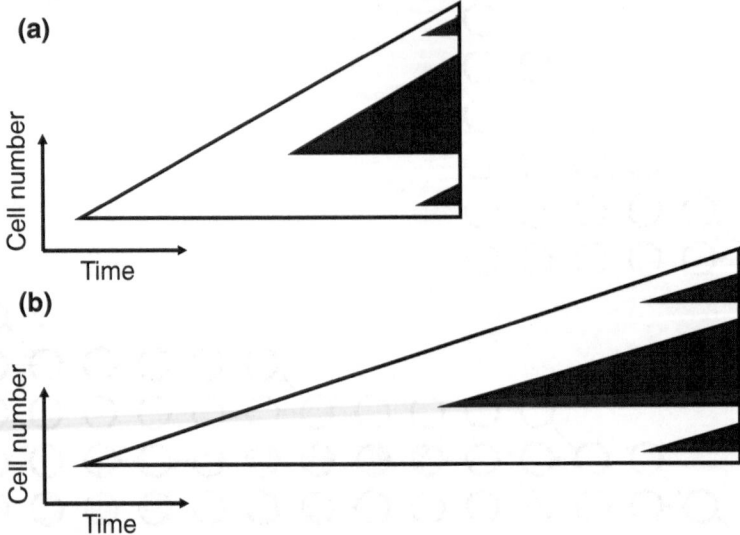

Fig. 6.5 The expected number of one-hit mutants does not depend on the presence of quiescence. (**a**) represents a colony with no quiescence, and there is quiescence in (**b**). The white triangles depict growing colonies of cells (cells with quiescence grow slower). The end size is the same in both cases. Dark triangles represent growing mutant clones inside the colonies. The total number of mutant colonies is the same in both cases (the same number of cell divisions). The mutant colonies in (**b**) have a longer time to grow, but at the same time they grow slower. Therefore the resulting frequency of mutants is the same in (**a**) and (**b**)

resistance to two drugs; higher numbers of drugs can be treated similarly. We build our arguments as follows.

6.5.2 The Number of Cycling One-Hit Mutants is Independent of the Quiescence Parameters

Cycling mutants are produced by cycling wild-type cells and they grow according to the same law as the cells producing them. When α increases (or β decreases), the mutant clones grow more slowly because of quiescence, but at the same time they have more time to grow, see Fig. 6.5. In other words, the changes in the mutant growth are completely compensated by the change in the time of growth. Therefore, we conclude that the number of cycling 1-hit mutants in a colony of a given size is also independent of quiescence.

6.5.3 The More Quiescence There is in the Colony, the Smaller is the Total Number of Cycling Wild-Type Cells

This result is actually a consequence of a more general statement, that for each cell type (that is, cells resistant to 0, 1, 2 etc drugs), the number of quiescent cells divided by the number of cycling cells is given by $\alpha/(L - \alpha)$, see Sect. 6.3.3. The particular fact that we will need is that, up to a small correction, the number of quiescent wild-type cells in a colony of size N is given by $\alpha N/L$, whereas the number of cycling wild-type cells is given by $(1-\alpha/L)N$ (here we assume that the mutation rate is small compared to 1, which is a safe bet). That is, the number of cycling wild-type cells decreases with α.

6.5.4 The Probability of Two-Drug Treatment Failure (Due to Resistance) Increases with the Quiescence Rate

Our calculations show that the probability of treatment failure, caused by resistant mutants, rises with the level of quiescence in the context of therapy with two separate

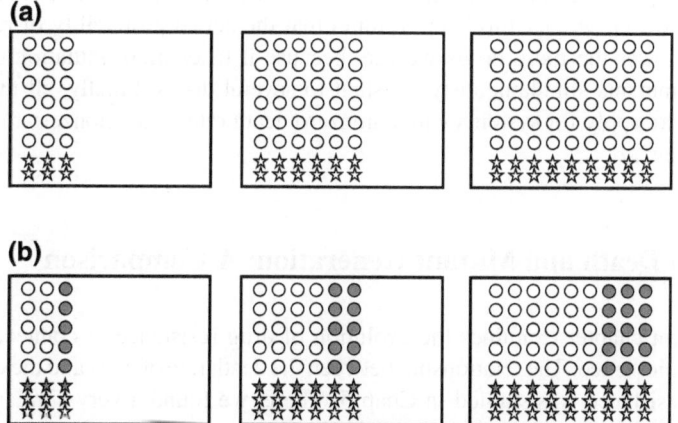

Fig. 6.6 A schematic illustrating the argument stating that the probability to produce 2-hit mutants increases with quiescence. Each rectangle represents a colony of cells. There are three moments of time shown, first we have $N = 24$, then $N = 48$ and finally $N = 72$. Circles represent wild-type cells, and stars—one-hit mutants. Gray shading denotes the state of quiescence for wild-type and mutant cells. In (**a**), we assume no quiescence ($\alpha = 0$), whereas in (**b**) there is a probability to become quiescent (with $\alpha = 1/3$). The number of cycling 1-hit mutants (empty stars) is the same in (**a**) and (**b**) for the same values of N. The number of quiescent wild-type cells is given by the fraction α of all wild-type cells (e.g. 1/3 in (**b**)). At each moment of time, one of the cycling cells is picked for reproduction. We can see that the probability to pick a 1-hit mutant is always higher in (**b**) than in (**a**), because the fraction of cycling one-hit mutants increases as the tumor grows. Therefore the probability to create a 2-hit mutant is higher in (**b**)

drugs (Fig. 6.3b–d). This is a direct consequence of the previous two sections. Let us consider a colony consisting of wild-type and 1-hit mutant cells. Let us "watch" the colony grow by tracking each of N-1 cell divisions, see Fig. 6.6. Whenever a cell division happens, it may be a division of a cycling wild-type cell, or a division of a cycling 1-hit mutant cell. It is only the latter process which in principle may lead to the generation of two-drug resistance. The probability to create a double mutant at each division is proportional to the probability that a 1-hit mutant (and not a wild-type) cell divides. The number of cycling wild-type cells in a colony of a given size is a decreasing function of α, whereas the number of cycling 1-hit mutants is independent of α (see the two previous paragraphs). Therefore, as α increases, the relative abundance of cycling 1-hit mutants increases. In other words, among all cycling cells, the percentage of mutants increases with α, and so does the probability to create 2-hit mutants. Thus, the probability of resistance generation against two drugs increases with quiescence parameters.

6.5.5 Generalizations

These results can be generalized. First of all, we can show by similar methods that the probability of mutant generation increases with quiescence for 3- and higher-degree mutants (Fig. 6.3). In fact, the dependence becomes stronger for larger numbers of drugs. However, we need to keep in mind that the actual probability of resistance becomes lower the more drugs we use, because it takes more mutation events to generate mutants simultaneously resistant to several drugs. Finally, all the results derived here apply for systems with a nonzero death rate, and a nonzero rate of cell "awakening", β.

6.6 Cell Death and Mutant Generation: A Comparison

The current chapter examines the evolution of drug resistance in connection with cellular quiescence. The relationship between the death rate of cells and the evolution of drug resistance was studied in Chap. 5. There, we found a very similar pattern. The probability of treatment failure was independent of the death rate of tumor cells in the context of therapy with a single drug. However, when treatment was assumed to occur with two or more drugs, the probability of treatment success depended on the death rate of tumor cells. The higher the death rate of tumor cells relative to their division rate, the higher the probability that mutant cells that are resistant against all drugs induce failure of therapy. While this result is identical to that observed for cellular quiescence, the reason for it is different. It is explained in the remaining part of this section.

6.6.1 The Probability of Pre-Existence of One-Hit Resistant Mutants is Independent of the Death Rate

As explained in Chap. 5, the probability of creating resistance before the start of treatment is defined by the probability to have at least one 1-hit mutant at a given colony size, which is given by *(probability to create a mutant)* × *(probability for a mutant clone to survive)*. The probability to create a mutant clone is proportional to the number of cell divisions. In turn, the number of cell divisions is a changing function of the death rate. With a zero death rate it takes exactly N-1 cell divisions to go from 1 cell to N cells. As the death rate increases, it can take a lot more cell divisions to expand, because cell divisions are (partially) countered by cell deaths. Therefore, there are more cell divisions for a larger death rate, and as a consequence, more 1-hit mutants are produced. However, the probability for a mutant to survive is a decreasing function of the death rate, which compensates the gain in the number of clones produced. Therefore the probability to create resistance against one drug is independent of the death rate.

It is interesting to note that the number of one-hit mutants is a growing function of both the death rate and the senescence rate, but for different reasons. If we increase the death rate, the total number of cell divisions to reach size N will increase, and so will the number of mutants (but the average size of a clone size will remain the same). If we increase α, the total number of divisions will not change but the average clone size will grow, again leading to an increase in the total mutant number.

6.6.2 The Probability of Pre-Existence of Two-Hit Resistant Mutants Increases with the Death Rate

While the probability to have 1-hit mutants is independent of the death rate, the average number of 1-hit mutants that are produced and survive by the time the tumor reaches size N is an increasing function of the death rate. The reason is as follows. The mutants are produced more often at higher death rates (because of the increased total number of cell divisions). Thus, more mutants are seeded to undergo clonal expansion. However, the size of the mutant clones is independent of the death rate (in the same manner as it was independent of the quiescence parameters, see Fig. 6.3). Therefore the total amount of 1-hit mutants present at size N is an increasing function of the death rate. As a direct consequence of this, the probability to have 2-hit mutants at size N is also an increasing function of the death rate. This explains why the likelihood of 2-drug resistance is a growing function of cell death. This result can be extended to a larger number of drugs.

As we saw in this chapter, the probability of preexistence of two- (and more) hit mutants also increases with the amount of quiescence, but again, this happens for different reasons, see Sect. 6.5.4.

6.7 Summary

This chapter examined the effect of CML stem cell quiescence on the evolution of drug resistant mutants. The model showed that in the context of treatment with a single drug, parameters that determine the kinetics of cellular quiescence do not affect the probability of treatment failure as a result of drug resistant mutants. On the other hand, if two or more drugs are used in combination to treat the cancer, then treatment failure as a result of drug resistance is promoted by the occurrence of cellular quiescence. Interestingly, while cellular quiescence significantly prolongs the time until the cancer has dropped to low numbers or has been driven extinct, the model predicts that drug resistance does not evolve during the treatment phase in this case. Increased cellular quiescence, however, affects the pretreatment evolution of resistance; it increases the likelihood that resistant mutants are generated during the growth phase of the cancer before therapy is initiated. These results add to previous theoretical work that highlighted in a different context the role of tumor stem cells and stem cell status in treatment responses [1, 2].

Problems

6.1 Derive deterministic system (6.3–6.4) from the master Eq. (6.1).

6.2 Study the fixed points of system (6.8–6.9). Perform a linear stability analysis of the equilibria.

6.3 Write down the exact solution of system (6.11–6.15).

6.4 Derive expression (6.17) by using a double Taylor expansion of the solutions.

6.5 Derive a deterministic system similar to Eqs. (6.11–6.15), but with $d_s \neq 0$, and with $\alpha_s = \beta_s = 0$. Solve this system. These calculations illustrate the explanations presented in Sect. 6.6.

6.6 Numerical project

Modify the program obtained for numerical project 4.8 to include the effect of cellular quiescence. How does the probability of treatment success depend on the quiesnence parameters, α and β, for different numbers of drugs m?

References

1. Enderling, H., Park, D., Hlatky, L., Hahnfeldt, P.: The importance of spatial distribution of stemness and proliferation state in determining tumor radioresponse. Math. Model Nat. Phenom. **4**(3), 117–133 (2009)
2. Piotrowska, M. J., Enderling, H., van der Heiden, U., Mackey, M.C.: Mathematical Modeling of Stem Cells Related to Cancer, in Cancer and stem cells (eds. T. Dittmar & K.S. Zänker), Nova Science Publisher, Hauppauge, NY (2008)

Chapter 7
Combination Therapies: Short-Term Versus Long-Term Strategies

Abstract The previous chapters have shown that a combination of several drugs might be needed to overcome the problem of drug resistance in more advanced phases of the disease. At the same time, the basic kinetics of tumor cell decline indicates that treatment might have to be applied for a relatively long period of time. While the degree of side effects is relatively small with targeted drugs, the prolonged use of combination therapy could have a negative impact on the patients' wellbeing. Using mathematical models, this chapter shows that the number of drugs used in combination can be reduced over time as the tumor cell population declines, without compromising the chances of treatment success. It is shown that the cancer size at which the number of drugs can be reduced does not correlate with the two phases of CML decline. Neither does it depend much on kinetic parameters of CML growth, except for the mutation rates. This is practical because even without any information on most parameters, and using only the data on the cancer size and the mutation rate, it is possible to predict at which stage of treatment the number of drugs can be reduced. Dependence of the treatment strategy on the activity spectra and potency of the drugs is discussed.

Keywords Side-effects · Targeted drugs · Prolonged treatment · Cellular quiescence · Combination therapy · Number of drugs · Stochastic and deterministic modeling · Activity spectra of drugs · Potency of drugs · Drug specificity · Partially resistant mutants · Fully-resistant mutants · Probability of treatment success · Kinetic rates · Tumor size · Mutation rate

7.1 Introduction

The previous chapters have shown that the simultaneous combination of several drugs can prevent treatment failure due to drug resistance. The reason is that cell mutants that are resistant against several drugs are unlikely to exist in the tumor cell

N. L. Komarova and D. Wodarz, *Targeted Cancer Treatment in Silico*,
Modeling and Simulation in Science, Engineering and Technology,
DOI: 10.1007/978-1-4614-8301-4_7, © Springer Science+Business Media New York 2014

population. The data on basic treatment dynamics discussed in Chap. 3 indicate that CML therapy with targeted drugs is characterized by a bi-phasic decline pattern, with the second decline phase being significantly slower than the first phase. Hence, while in the absence of resistance, a complete remission can be expected, this is likely to take a relatively long time. Administration of multiple drugs is characterized by more toxicity than treatment with a single drug, and this could pose problems for combination treatments, especially during long-term therapy. How can we reduce the number of drugs administered to patients long term? The idea is as follows. While cancer cells die and decline during treatment, it is possible that some of the partially resistant mutants that existed before therapy go extinct. This applies especially to those mutants that existed at the lowest level, i.e., the mutants that are resistant to more than one drug. If such mutants go extinct during the course of treatment, then it should be possible to reduce the number of drugs after a certain period of time and continue long-term therapy with fewer drugs than in the beginning. For example, if three drugs are necessary to treat a cancer of a given size, mutants that are resistant to two drugs might have gone extinct once the cancer size has been reduced by several orders of magnitude. Hence, two drugs might be sufficient to continue therapy in the long term. These concepts are explored here in the context of mathematical models that are an extension of the model discussed in Chap. 6. We show how to calculate the tumor size at which it is safe to reduce the number of drugs during the treatment.

7.2 Methodology

We assume the following treatment strategy, Fig. 7.1. For time $0 < t < t_*$, there is no treatment. For time $t_* < t < t_{**}$, m noncross-resistant drugs are applied in combination (we refer to this stage as treatment stage I). Finally, for time $t > t_{**}$, the number of drugs given in combination is reduced. That is, starting from time t_{**}, there is an $\bar{m}$-drug treatment, with $\bar{m} < m$ (treatment stage II).

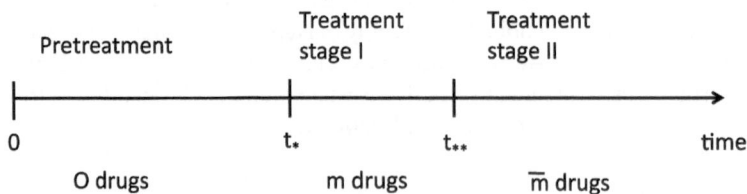

Fig. 7.1 The three phases: pretreatment, treatment stage I, and treatment stage II

7.2.1 Stochastic Methods

As explained in Chap. 4, all the variables can be separated into classes, such that in each class k, all the cell types are resistant to exactly k drugs and susceptible to $m - k$ drugs. Let us suppose that within each class, the birth, death, and other rates are equal. We will also assume that all mutation rates are equal to each other. Further, we will include the phenomenon of cellular quiescence in the model, as described in Chap. 6. The associated rates are also assumed to be symmetric. Then, for times $0 < t < t_{**}$, the equations for all types within the same class are identical, and we have:

$$\dot{\xi}_k = L_k(1 - (m - k)u)\xi_k^2 + [(m - k)L_k u \xi_{k+1} - (L_s + d_k + \alpha_k)]\xi_k + d_k + \alpha_k \eta_k,$$
$$\dot{\eta}_k = \beta_k(\xi_k - \eta_k).$$

For times $t > t_{**}$, within each class we will have different behavior. Indeed, let us suppose that $m = 3$ and $\bar{m} = 2$, and drugs 1 and 2, but not 3, are applied during the second stage of treatment, see Fig. 7.2. As before, in class $k = 3$ we will have one fully resistant type (111). In class $k = 2$, we have one fully resistant type (110) and two partially resistant types (101 and 011). Therefore, class $k = 2$ splits into two subclasses, resistant and susceptible. Next, we note that equations for class 1 depend on variables in class 2. Let us separate such equations in class 1 into two subclasses: the first subclass includes only the variables whose equations depend on resistant variables from class 2. The second subclass only depends on susceptible variables in class 2. We will call the former subclass resistant, and the latter one susceptible. In the general case where $\bar{m} < m - 1$, the number if subclasses

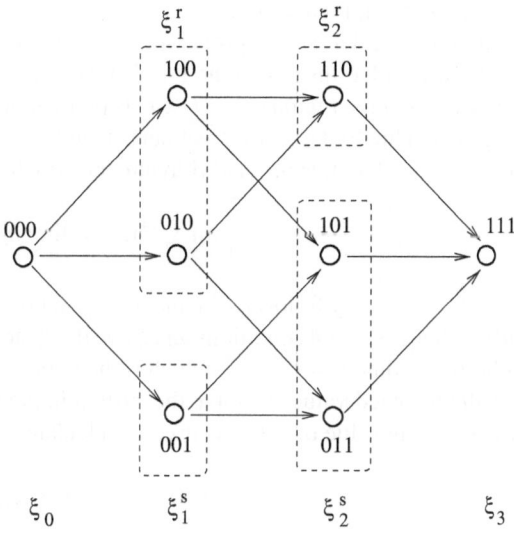

Fig. 7.2 Combinatorial mutation diagrams for different resistance classes. Treatment with $m = 3$ and then with $\bar{m} = m - 1 = 2$ drugs

within each class is bigger than two, and the description becomes quite complex. For the special case $\bar{m} = m - 1$, all classes below class m can be separated into two subclasses, depending on whether or not they depend on the fully resistant type of class $m - 1$. The corresponding variables will be denoted by subscripts "s" and "r" for susceptible and resistant. We have

$$\dot{\xi}_m = L_m \xi_m^2 - (L_m + d_m + \alpha_m)\xi_m + d_m + \alpha_m \eta_m,$$

$$\dot{\xi}_{m-1}^r = L_{m-1}(1 - u)(\xi_{m-1}^r)^2 + d_{m-1}^r + \alpha_{m-1}\eta_{m-1}^r$$
$$\quad + [uL_{m-1}\xi_m - (L_{m-1} + d_{m-1}^r + \alpha_{m-1})]\xi_{m-1}^r,$$

$$\dot{\xi}_{m-1}^s = L_{m-1}(1 - u)(\xi_{m-1}^r)^2 + d_{m-1}^s + \alpha_{m-1}\eta_{m-1}^s$$
$$\quad + [uL_{m-1}\xi_m - (L_{m-1} + d_{m-1}^s + \alpha_{m-1})]\xi_{m-1}^r,$$

$$\dot{\xi}_l^r = L_l(1 - (m - l)u)(\xi_l^r)^2 + d_l + \alpha_l \eta_l^r$$
$$\quad + [(m - 1 - l)uL_l\xi_{l+1}^r + uL_l\xi_{l+1}^s - (L_l + d_l + \alpha_l)]\xi_l^r, \quad 1 \le l \le m - 2$$

$$\dot{\xi}_l^s = L_l(1 - (m - l)u)(\xi_l^s)^2 + \alpha_l \eta_l^s$$
$$\quad + [(m - l)uL_l\xi_{l+1}^s - (L_l + d_l^r + \alpha_l)]\xi_l^s + d_l, \quad 1 \le l \le m - 2$$

$$\dot{\xi}_0 = L_0(1 - mu)\xi_0^2 + d_0 + \alpha_0 \eta_0$$
$$\quad + [(m - 1)uL_0\xi_1^r + uL_0\xi_1^s - (L_0 + d_0 + \alpha_0)]\xi_0,$$

with the following equations for the η's:

$$\dot{\eta}_l^{s,r} = \beta_l(\xi_l^{s,r} - \eta_l^{s,r}), \quad 0 \le l \le m.$$

In the above equations, we denoted by $d_{m-1}^{s/r}$ the death rate of susceptible/resistant cells in class $m - 1$. If we use less than $m - 1$ drugs, then the simplification of resistance (sub)-classes becomes less useful and we should use the full asymmetrical description with binary indices.

In order to calculate the probability of treatment success in a two-stage treatment, we need to solve the equations for characteristics in three times. The time axes is divided into three regions: $0 < t < t_*$ is pretreatment, then $t_* < t < t_{**}$ is treatment stage I, and finally $t_{**} < t$ is treatment stage II. First, we need to solve the equations for the characteristics in stage II treatment, which is equivalent to finding the limit

$$\lim_{t \to \infty} \xi_l^{r/s}(t) = \lim_{t \to \infty} \eta_l^{r/s}(t)$$

(the above equality follows from the fixed point of the equations for η_l). Then we use these values as initial conditions and solve the system for stage I treatment, to find the solutions at time $t = t_{**} - t_*$, which is the duration of stage I treatment. Finally, we use the obtained values to solve the system in pretreatment for the duration $t = t_*$. The resulting value of $\xi_0(t_*)$ is used to calculate

$$\lim_{t \to \infty} \Psi(0, \ldots, 0; t) = (\xi_0(t_*))^{M_0}.$$

Let us use the usual symmetry assumptions,

$$L = L_i, \quad \alpha = \alpha_i, \quad \beta = \beta_i, \quad 0 \le i \le m.$$

and further assume that if a cell is susceptible to at least one of the m drugs, it dies with the same rate as a fully susceptible cell. Let us suppose that $\bar{m} = m - 1$. In this case, we can solve the system in treatment stage II analytically. We obtain in the limit $t \to \infty$, to the highest order in small u[1]:

$$\xi_m = \frac{D}{L},$$

$$\xi_l^r = 1 - \frac{(m - l - 1)!(L - D)l^{m-l-2}u^{m-l-1}}{(D + H - L)^{m-l-1}},$$

$$\xi_l^s = 1 - \frac{(m - l)!(L - D)l^{m-l-1}u^{m-l}}{(D + H - L)^{m-l}}.$$

In particular, we have

$$\xi_0 = 1 - \frac{(m - 1)!(L - D)l^{m-2}u^{m-1}}{(D + H - L)^{m-1}}. \tag{7.1}$$

Note that this expression is similar to the case where the treatment involves m drugs; therefore the formula is obtained from the above expression by replacing $m - 1$ with m.

It is clear that in the limit $t_{**} \to \infty$, the two-stage treatment is equivalent to a one-stage treatment with m drugs. Similarly, if $t_{**} \to t_*$, then the two-stage treatment is equivalent to a one-stage treatment with $m - 1$ drugs, see Fig. 7.1.

The methodology described here allows us to calculate the probability of treatment success as a function of time t_{**}, or equivalently, the tumor size N_{off}, which corresponds to the start of stage II treatment. The goal is to find the largest size N_{off} still compatible with the probability of treatment success being within a small margin (say, δ), of the target probability of treatment success (the probability achieved with a one-stage treatment with m drugs). The results obtained by this method are presented later in this chapter, see Fig. 7.6.

7.2.2 Deterministic Methods

The growth stage (before treatment) can be described by the usual ordinary differential equations :

---

[1] In all the numerical calculations we use the exact limiting values, and not their truncation in terms of u presented above.

$$\dot{x}_i = (m - i + 1)uLx_{i-1} + [L(1 - (m - i)u) - D - \alpha]x_i + \beta y_i, \quad 1 \le i \le m;$$
$$\dot{x}_0 = [L(1 - mu) - D - \alpha]x_0 + \beta y_0,$$
$$\dot{y}_i = \alpha x_i - \beta y_i, \quad 0 \le i \le m,$$

where $x_i(t)$ and $y_i(t)$ are the expected numbers of cycling and quiescent cells in each class. The initial conditions are

$$x_i(0) = 0, \quad 1 \le i \le m, \quad x_0(0) = 1, \quad y_i(0) = 0 \ \forall i.$$

The solution of this system must be evaluated at time t_* given by $\sum_{i=0}^{m}(x_i(t_*) + y_i(t_*)) = N$. We denote $x_i(t_*) \equiv x_i^0$, $y_i(t_*) \equiv y_i^0$ for $0 \le i \le m$. Next, the treatment stage with m drugs is described by a similar system, only the death rate, d, is given by $D + H$ for all types except the fully resistant type, x_m, where we have $d = D$ as before. The initial conditions for the treatment stage are given by

$$x_i(0) = x_i^0, \quad y_i(0) = y_i^0 \ \forall i.$$

In order to estimate the size at which one drug can be removed, let us determine the average number of mutants resistant to $m - 1$ drugs, and find the time when this quantity becomes less than one. In other words, we solve the equation for t_{**}, $x_{m-1}(t_{**}) + y_{m-1}(t_{**}) = 1$, and then evaluate

$$N_{\text{off}}^{\text{det}} = \sum_{i=0}^{m}(x_i(t_{**}) + y_i(t_{**})). \tag{7.2}$$

This method is illustrated in Fig. 7.3, where the calculation of t_{**} and $N_{\text{off}}^{\text{det}}$ is demonstrated. We can see that the shape of the function $x_{m-1}(t) + y_{m-1}(t)$, as well as the total population size, has a characteristic form of biphasic decline typical for populations with quiescent cells, see Chap. 3. Note that the location of the "knee" of the decline is the same for all cell subpopulations.

The quantity $N_{\text{off}}^{\text{det}}$ calculated as shown above is usually slightly larger than the value predicted by the stochastic theory, but it gives the correct general location of the drop in the curves for the probability of treatment success as functions of N_{off}. Figure 7.6 shows the location of $N_{\text{off}}^{\text{det}}$ for the particular parameter set used in that case study.

We have seen that the quantity $N_{\text{off}}^{\text{det}}$ strongly depends on the mutation rate, u, and is not sensitive to changes in N. Let us investigate how it depends on other parameters. We start with the parameter α.

Figure 7.4a presents the quantity $x_{m-1} + y_{m-1}$, the expected number of mutants resistant to $m - 1$ drugs, as a function of time, for varying values of the parameter α. The time t_{**} at which one drug can be removed is determined by $x_{m-1} + y_{m-1} = 1$. We can see that as α grows (more quiescence in the system), the time t_{**} grows. That is, the number of drugs can be reduced after a longer span of time for systems with

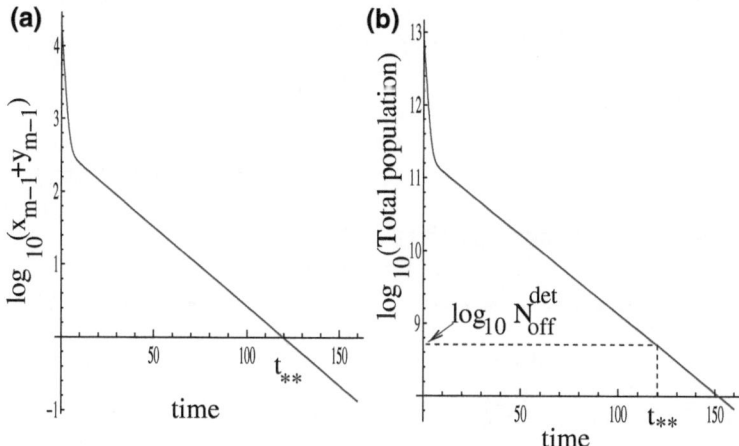

Fig. 7.3 A deterministic method of estimating the tumor size at which the number of drugs can be reduced. (**a**) The expected number of mutants resistant to $m - 1$ drugs is plotted as a function of time; we use this to the time t_{**} where the curve crosses zero. (**b**) The total population size is plotted as a function of time. The value corresponding to t_{**} gives the tumor size where it is safe to reduce the number of drugs by one. $N = 10^{13}$, $L = 1$, $D = 0$, $H = 2$, $\alpha = 0.01$, $\beta = 0.05$, $M_0 = 100$

Table 7.1 The dependence of the size $N_{\text{off}}^{\text{det}}$ on parameter values of the system: (a) α, (b) β and (c) d, see Fig. 7.4a–c

(a)	α	1	10^{-1}	10^{-2}	10^{-3}	10^{-4}	10^{-5}
	$\log_{10} N_{\text{off}}^{\text{det}}$	7.8	8.6	8.7	8.7	8.7	8.5
(b)	β	1	10^{-1}	10^{-2}	10^{-3}	10^{-4}	10^{-5}
	$\log_{10} N_{\text{off}}^{\text{det}}$	8.4	8.5	8.7	8.7	8.7	8.7
(c)	d	0	0.1	0.2	0.5	0.4	0.5
	$\log_{10} N_{\text{off}}^{\text{det}}$	8.7	8.6	8.5	8.4	8.3	8.2

The rest of the parameters are as in Fig. 7.3

higher quiescence. However, the tumor *size*, $N_{\text{off}}^{\text{det}}$, at which the drug reduction can take place, remains virtually unchanged with α, see Table 7.1(a).

An interesting observation is that the time t_{**} (the points where the curves in Fig. 7.4 cross zero) can correspond to the first, fast, or the second, slow, phase of decay in the biphasic decline of the tumor (Chap. 3). In the particular example of Fig. 7.4a, the regime corresponding to $\alpha = 10^{-5}$ is characterized by the time t_{**} still in the first phase of decline. All larger values of α correspond to time t_{**} in the second phase. This means that the location of the "knee" of the decline is not indicative of when a drug can be removed. For some parameters, a drug can be taken away while still in the first phase of the decline, and for others one has to wait for a long time, well into the second phase, until a drug can be taken off.

Figure 7.4b presents the dependence of the time until a drug can be taken off, on the parameter β. Similarly to the previous case, the more quiescence there is in

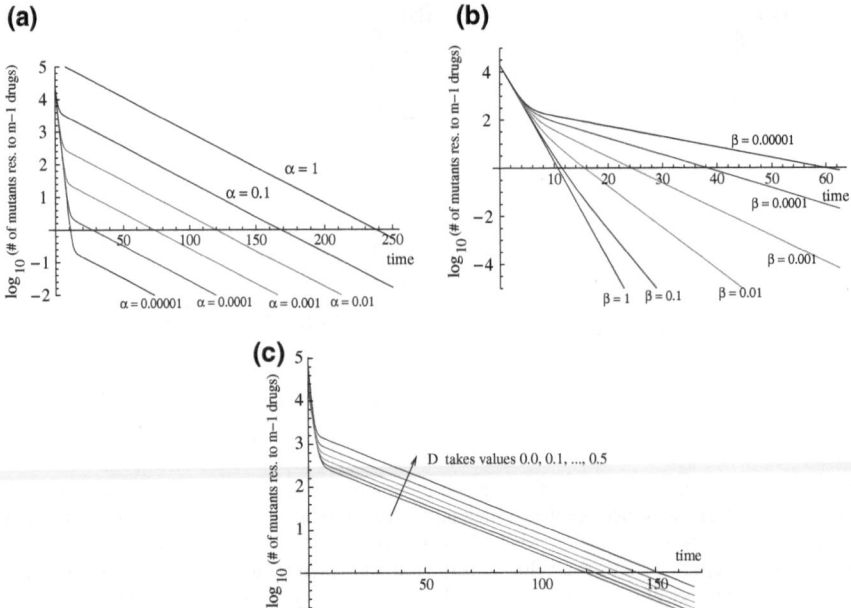

Fig. 7.4 The expected number of mutants resistant to $m-1$ drugs, as a function of time, for varying values of the parameter **(a)** α, **(b)** β and **(c)** d. $N = 10^{13}$, and the rest of the parameters are as in Fig. 7.3. The sizes $\log_{10} N_{\text{off}}^{\text{det}}$ calculated for each of the values of α, β and d are given in Table 7.1

the system (smaller values of β), the longer one needs to wait until a drug can be removed. But at the same time, the size corresponding to taking off a drug, $N_{\text{off}}^{\text{det}}$, is largely independent on β, as it was on α (see Table 7.1(b)).

Figure 7.4c shows a similar study with the value d changing. We can see that the time t_{**} grows with d, and the size $N_{\text{off}}^{\text{det}}$ is only weakly dependent on d (Table 7.1(c)).

7.3 The Models

We use three different methods of calculating the tumor size where it becomes safe to decrease the number of drugs by one. These are the full stochastic method, the ordinary differential equations for the expected values, and the "number of drugs" diagram which is explained in the next section, see Fig. 7.5.

Our stochastic model describes the dynamics of the cancer cell population during growth and during treatment, see also [1–4]. We distinguish between active, cycling cells, and quiescent cells. These correspond to the primitive CML cell populations that are thought to drive and maintain the disease. As in Chaps. 5, 6, active cells divide with a rate L, and die with a rate d. They can enter quiescence with a rate α. Quiescent cells re-activate with a rate β. The model describes treatment with m

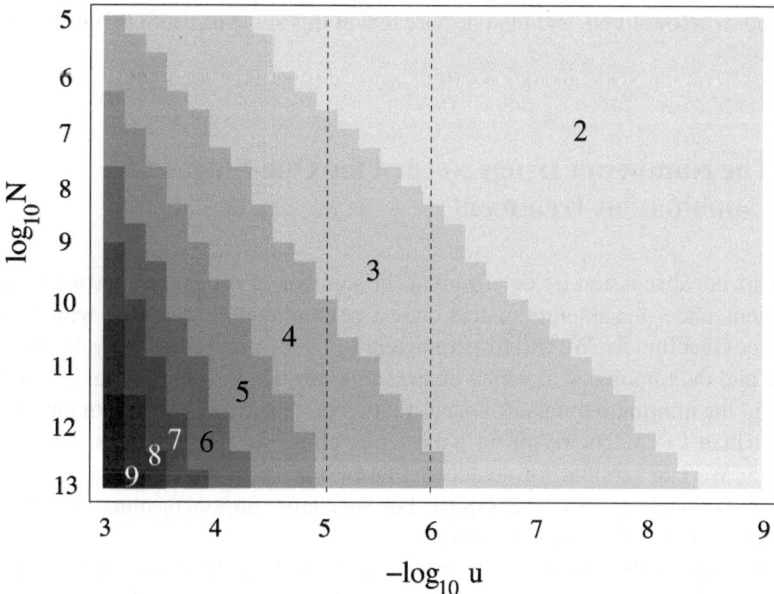

Fig. 7.5 The number of drugs needed for the probability of treatment success $1 - \delta$ with $\delta = 0.01$, as a function of u and N. Other parameters are: $L = 1$, $D = 0$, $H = 2$, $\alpha = 0.01$, $\beta = 0.05$, $M_0 = 100$. The jagged appearance of the boundaries between the domains with different values of m is due to a discrete method used to calculate it, and can be smoothed by using a simulation with a higher resolution

drugs, which affects the model coefficients. That is, the overall cell death rate is given by $d = D + H$, where D is the natural and H is the drug-induced death rate.

All cells are subdivided into resistance classes, ranging from fully susceptible to fully resistant. A cell can acquire resistance to a given drug through a mutation with a rate u. We assume that the cancer grows in the absence of treatment ($L > D$) until it has reached N cells. A one-stage treatment is a combination therapy with m drugs (to prevent resistance), leading to an exponential decline of all the susceptible cells ($d = D + H > L$). Further, we consider two-stage treatment strategies, where during stage I, m drugs are applied in combination, and then during stage II treatment, the number of drugs is reduced.

The stochastic analysis allows us to calculate the probability of treatment success (related to the extinction probability for cells), given the model parameters. We use these methods to calculate the tumor size, at which the number of drugs in a combination treatment can be reduced.

A simpler, approximate method of estimating the size at which the number of drugs can be safely reduced is provided by deterministic modeling. There, we track the dynamics of the average population sizes for different resistance classes under stage I therapy, and specifically calculate the expected number of mutants resistant to $m - 1$ drugs. We determine the time when the expected size of this population is reduced to below 1 cell, and then declare that at this stage it is safe to remove one of the drugs.

7.4 The Number of Drugs Needed for One-Stage Combination Treatment

We start our discussion by determining the number of drugs, m, that we need for treatment, under the assumption that one-stage combination treatment with m drugs continues indefinitely. We will fix parameters L, D, H, α and β, and vary the mutation rate u and the tumor size at which treatment is started, N. For each pair (u, N), we can find the minimum number of drugs that gives a probability of treatment success greater than $1 - \delta$. The result for a particular choice of parameters is presented in Fig. 7.5. We can see that for smaller mutation rates and smaller tumor sizes, two drugs are enough to treat successfully. For very large tumors or mutation rates, the number of drugs grows significantly.

The value of the mutation rate plays an important role in this analysis, and it strongly influences the number of drugs needed for treatment. In Chap. 5, we argued that most of the time, resistance preexists treatment, that is, a number of resistant mutants are already present at the start of therapy. The same holds in the present investigation; the mutation rates are of utmost importance because it is by the way of mutations that the resistant clones come about. In our mathematical approach we consider several stages of tumor evolution, the first one being "before treatment". According to the model, it is during that first stage that resistance is mostly generated by mutations and may cause treatment failure at later stages.

7.5 The Size at Which the Number of Drugs can be Reduced

Next, we assume that at some point during the treatment, the number of drugs is reduced. We will examine whether fewer drugs are sufficient to treat the cancer in the long term, once the number of cells has shrunk. Let us calculate the tumor size at which one drug can be taken away, such that the probability of treatment success does not decrease by more than δ. An important quantity is N_{off}, the tumor size at which treatment with m drugs is replaced with a treatment with $m - 1$ of the drugs. Figure 7.6 presents the probability of treatment success as a function of N_{off}, for $m = 3$ (that is, starting with a 3-drug treatment and reducing it to a 2-drug treatment). This is plotted for different values of N, the tumor size at which treatment is started. We observe the following interesting trends.

The probability of treatment success is flat for small values of N_{off}, and it is equal to the probability of treatment success for $m = 3$ drugs. In other words, if we remove one of the drugs when the number of cancerous cells is very low, it does not change

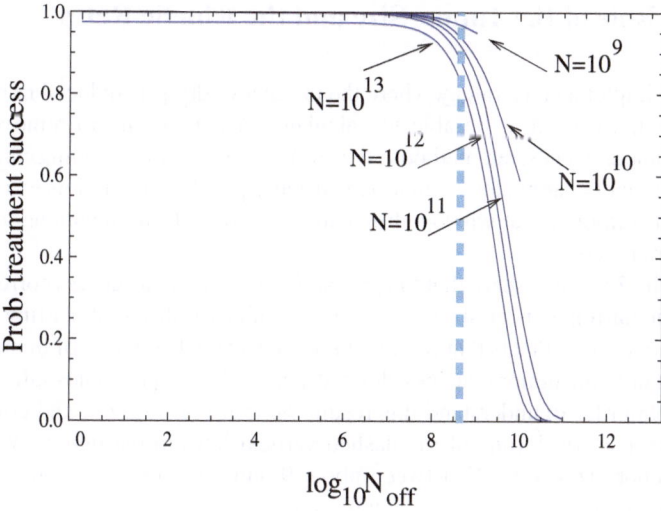

Fig. 7.6 Probability of treatment success as a function of N_{off}, the size at which the number of drugs is reduced by one. Different curves correspond to different treatment sizes, N. The initial number of drugs is $m = 3$, and $u = 10^{-6}$. The rest of the parameters are as in Fig. 7.5. The *thick dashed vertical line* corresponds to the quantity $N_{\text{off}}^{\text{det}}$, the predicted tumor size where it becomes safe to reduce the number of drugs

the probability of treatment success. Note that in the particular example of Fig. 7.6, the probability of treatment success for low N_{off} is very close (closer than δ) to 1 for all curves except for the one with $N = 10^{13}$. If $N = 10^{13}$ and $u = 10^{-6}$, we can see from Fig. 7.5 that the number of drugs necessary for a success rate greater than $1 - \delta$ is $m = 4$, that is, if $m = 3$, the probability of success is smaller than $1 - \delta$. This can be seen in the limiting value of the probability of success rate as $N_{\text{off}} \to 10$ for $N = 10^{13}$ in Fig. 7.6.

On the other hand, if we take away one drug too early (N_{off} near N), then the probability of treatment success is lower than that with m drugs throughout the course of therapy. There is a relatively sharp transition between maximum possible probability of treatment success, and a low probability of treatment success.

The main finding is that if we wait sufficiently long, the number of drugs used in combination treatment can be safely reduced by one. Using similar methods, we can show that for some parameter combinations, there are conditions when more than one drug can be safely removed from treatment. The tumor size at which this can be done can be calculated in similar ways.

7.6 The Role of the Tumor Size and the Kinetic Rates

In order to implement a strategy where the number of drugs is reduced in the course of treatment, we need to be able to calculate *when* it is safe to remove a drug. Given parameter values, our methods allow us to perform corresponding calculations. However, reliable parameter estimation is usually problematic. Let us examine how strongly the tumor size at which a drug can be removed from treatment depends on system parameters.

From Fig. 7.6, we observe that the probability of treatment success corresponding to different starting cancer sizes, N, is very similar for different starting sizes, N. One consequence of that is that for any tumor size at the beginning of treatment, one has to wait until the tumor reaches a defined size, and then it becomes safe to remove one drug. In order to understand this result, we return to Fig. 7.5, and consider the case where $u = 10^{-6}$ (one of the dashed vertical lines in the figure). We can see that for tumor sizes $\log_{10} N$ between about 9 and 12, three drugs are needed for the desired degree of success. For smaller tumor sizes, two drugs are enough. This simple reasoning explains intuitively why for any starting treatment size within this range, a drug can be removed from treatment at about size $N_{\text{off}} \sim 10^9$ for $u = 10^{-6}$. One needs to wait until the tumor size is reduced enough such that $m - 1$ drugs can accomplish the job successfully, and then take one drug off.

The above argument is not very precise, because Fig. 7.5 has been created assuming that for each pair (N, u), the tumor grows to size N and is subsequently treated with m drugs. The composition of a tumor that grew to size N_1 before start of treatment may be quite different from the composition of a tumor that was treated starting from $N_2 > N_1$ with m drugs and thus reduced to size N_1. However, we can use the chart in Fig. 7.5 for a rough estimate of reasonable sizes.

We have seen that the size at which the number of drugs can be safely reduced, strongly depends on the mutation rate, u, and is not sensitive to changes in N. It turns out that this quantity only weakly depends on other system parameters (see the Methodology section for details). Here is a summary of the results:

- For systems with higher rate of quiescence, α, one has to wait longer before it is safe to reduce the number of drugs. However, the tumor *size* at which the drug reduction can take place, remains virtually unchanged with α.
- The smaller values of β are (more quiescence in the system), the longer one needs to wait until a drug can be removed. But again, the size corresponding to taking off a drug is largely independent from β.
- The time lapse before it is safe to remove a drug grows with D, but the tumor size is only weakly dependent on the death rate.

We conclude that even though the time of drug removal is sensitive to exact parameter estimations, the *size* at which the number of drugs can be reduced is only weakly dependent on parameter values.

Finally, we investigate how the size at which a drug can be removed depends on the location of the "knee" of the biphasic tumor decline (Chap. 3). It turns out that the position of the "knee" is not indicative of when a drug can be removed. For some parameters, a drug can be taken away while still in the first, fast, phase of the decline, and for others one has to wait for a long time, well into the second phase, until a drug can be taken off.

7.7 Drugs with Different Specificity and Potency

In this section, we discuss some applications of our method to scenarios where the drugs have different characteristics. In particular, we will concentrate on the drugs' activity spectrum and potency.

We will say that a drug is characterized by a *broad activity spectrum* if it is able to bind to a large variety of mutant Bcr-Abl proteins, with a subsequent successful destruction of the cell. On the other hand, a drug with a *narrow activity spectrum* is a more *specific* agent, which is active against a relatively small number of Bcr-Abl variants. In terms of our modeling approach, the drug with a broader activity spectrum will be characterized by a smaller mutation rate, u, with which mutants resistant to the drug are generated. The more specific, or narrow, drug, is characterized by a larger mutation rate associated with the generation of mutants resistant to the drug.

By *potency*, we mean how effectively a drug kills cells which are susceptible to the drug; this is reflected in the drug-induced cell death rate characteristic of each drug, where higher potency is correlated with higher values of H.

Consider a two-drug treatment with drugs 1 and 2. Suppose that drug 2 is characterized by a broader activity compared to drug 1, and drug 1 is more specific. This setup is illustrated in Fig. 7.7, where a mutation diagram for all the resistance types is shown. Let us envisage a schedule whereby the tumor is first treated with a combination of drugs 1 and 2, and then the number of drugs is reduced by one. Which drug should remain and which one should be taken off?

First we assume that the potency of both drugs is the same, that is, $H_1 = H_2$. In order to identify the more effective treatment strategy in the presence of two drugs

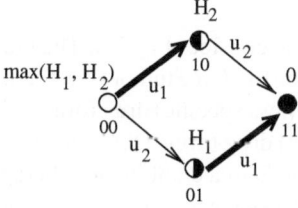

Fig. 7.7 Mutation diagram for a two-drug case where drug 1 has a narrower specificity (a larger mutation rate u_1). Cell types are denoted by *circles*: Fully susceptible cells (combinatorial type *00*) are denoted by an *empty circle*; partially resistant cells (*10* and *01*) are denoted by *half-shaded circles*; and fully resistant mutants (*11*) are *black circles*. The mutation rates and drug-induced death rates are marked. We have $u_1 > u_2$ (a *thicker arrow*)

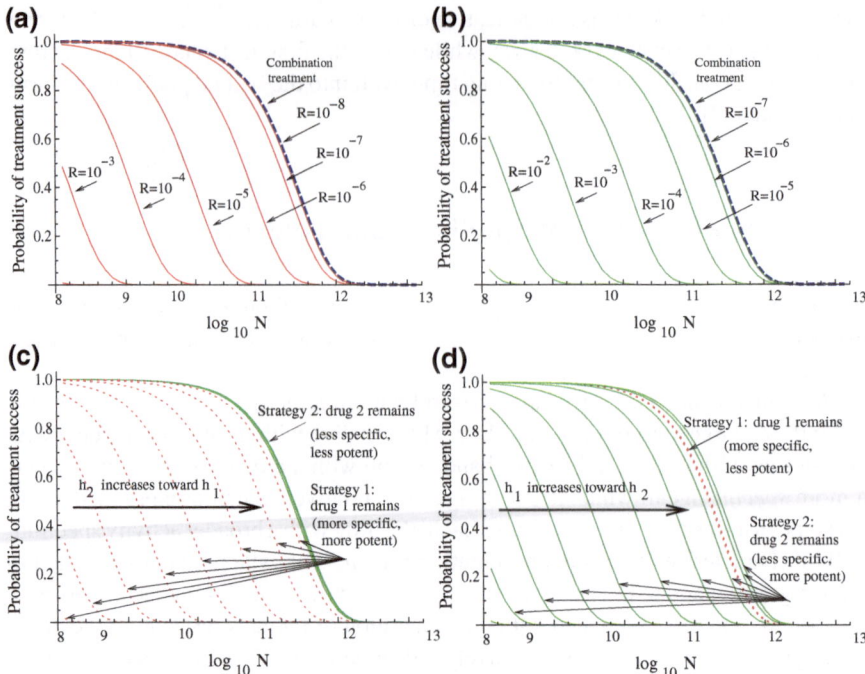

Fig. 7.8 Drugs of different specificity profiles: the probability of treatment success as a function of N, the tumor size, for different treatment schedules. Drug 1 always has a narrower specificity than drug 2. In **(a)-(b)** the two drugs have equal potency ($H/L = 5$). The *dashed thick line* corresponds to the combination treatment of two drugs. The other lines correspond to treatment schedules where one of the drugs is removed once the tumor size is reduced by a factor R, indicated in the figure. **(a)** Drug 1 is removed; **(b)** drug 2 is removed. In **(c)-(d)**, the two drugs have different potency. *Dotted red lines* correspond to "Strategy 1", where after a combination treatment, only drug 1 remains. *Solid green lines* correspond to "Strategy 2", where only drug 2 remains. **(c)** Drug 1 has a larger potency than drug 2; different lines correspond to the values of $H_2/L \in \{5, 10, \ldots, 50\}$, such that $H_2 \leq H_1 = 50L$. **(d)** Drug 1 has a smaller potency than drug 2; different lines correspond to values of $H_1/L \in \{5, 10, \ldots, 50\}$, such that $H_1 \leq H_2 = 50L$. The parameters are $D/L = 0.2$, $u_1 = 10^{-7}$, $u_2 = 10^{-6}$, $M_0 = 10^2$

with different activity spectra, consider Fig. 7.8. The graphs show the probability of treatment success as a function of N, the tumor size, for different treatment schedules. In Fig. 7.8a, the more active (less specific) drug (drug 2) is removed, and in Fig. 7.8b, the less active (more specific) drug (drug 1) is removed. The rightmost dashed thick line in both graphs corresponds to a combination therapy. The other lines show the probability of treatment success when a drug is discontinued; the drug is removed once the tumor is reduced by a factor R, which appears above each line. The goal is to keep the probability of treatment success as high as possible (the thick dashed line). We can see that in both cases, if the reduction happens sufficiently late in treatment

Table 7.2 A summary of the dynamics of different resistant types during the pretreatment stage and two stages of treatment

Stage	Mutants resistant to drug 1	Mutants resistant to drug 2
Pretreatment	Generated at rate u_1	Generated at rate u_2
Treatment stage I	Killed with rate H_2	Killed with rate H_1
Treatment stage II (drug 1 only)	Not killed	Killed with rate H_1
Treatment stage II (drug 2 only)	Killed with rate H_2	Not killed

(low values of R), the probability of treatment success will be unchanged. This is consistent with the results of the previous subsections.

We can see, however, that if the less specific drug is removed (Fig. 7.8a), then we need to wait until the tumor size is reduced by a factor of $R = 10^{-8}$; otherwise the probability of treatment success is significantly lower compared to that of continuous combination therapy. On the other hand, if the more specific drug is removed (Fig. 7.8b), then we can withdraw the drug earlier in treatment (more precisely, when the tumor size is reduced by a factor of $R = 10^{-7}$). This suggests that a more effective strategy is the one when the more specific drug is removed (Fig. 7.8b).

This result can be explained in the following way (see Table 7.2). Because of the differential activity spectra of the two drugs, the resistant mutants are accumulated at different rates before treatment. Mutants resistant to drug 1 are generated at a higher rate because $u_1 > u_2$. As a result, by the time treatment starts, there tends to be more mutants resistant to drug 1. During treatment stage I, both types of partially resistant mutants are killed at rate $H_1 = H_2$, such that by the time one of the drugs is removed (beginning of stage II treatment), there will be more mutants resistant to drug 1. To maximize treatment success, these mutants have to be removed; they are only susceptible to drug 2 (and resistant to drug 1). Therefore, it is more effective to continue treating with drug 2 to which these mutants are susceptible.

Next, let us add a layer of complexity and assume that the two drugs can differ not just in their specificity, but also in their potency. In terms of our analysis this means that the drug-induced cell death rates, H_1 and H_2, are different. Below we consider the two cases, $H_1 > H_2$ and $H_2 > H_1$, see Fig. 7.8c, d.

7.7.1 The Drug with a Narrower Activity is More Potent

We consider the situation where $H_1 > H_2$. As before, $u_1 > u_2$. In words, drug 1 is more specific but it is also more potent than drug 2. This case is similar to the case $H_1 = H_2$: the more effective treatment schedule involves continuing with the broader drug (drug 2) while removing the narrower drug (drug 1). The argument is as follows. By the time treatment starts, there are more mutants resistant to drug 1 than there are mutants resistant to drug 2 (see Table 7.2; as before, they are produced at a higher rate). During stage I treatment, the more abundant mutants are killed at a lower rate, H_2, such that at the beginning of stage II treatment, mutants resistant to drug 1 are still more prevalent and they pose the biggest risk. Therefore the drug applied in stage II treatment should be able to effectively kill these mutants. This is drug 2, as the more prevalent mutants are resistant to drug 1. The above ideas are illustrated in Fig. 7.8c, where we plot the probability of treatment success as a function of tumor size, for two strategies: "Strategy 1" when drug 1 continues (dotted red lines) and "Strategy 2" when drug 2 continues (solid green lines). We keep all the parameters fixed except varying the drug 1-induced cell death rate, H_1 (while it always remains below H_2). Note that for "Strategy 2", the solid green line which appears thick is made up of several lines that have very similar probabilities of treatment success and are hence close together. First of all we can see that it is always advantageous to keep drug 2 and remove drug 1 (the corresponding probabilities of success are always higher). Another interesting observation is the following. As we vary the strength of drug 2, the probability of treatment success of "Strategy 1" changes dramatically, while this has almost no effect on "Strategy 2". This is because for "Strategy 1", the mutants resistant to drug 1 are only killed during stage I treatment by drug 2, thus its strength is of a crucial importance. Under "Strategy 2", these mutants are continued to be exponentially killed throughout treatment.

7.7.2 The Drug with a Narrower Activity is Less Potent

If $H_1 < H_2$, we can say that in this case drug 1 is inferior to drug 2 because it is more specific and less potent. Under this scenario, interestingly, the choice of strategy will depend on the extent to which drug 2 is more potent than drug 1. During treatment stage I, the more abundant mutants (the ones resistant to drug 1) will be killed at a higher rate (H_2) than the less abundant mutants resistant to drug 2. Therefore, if by the time stage II treatment starts, there are still more mutants resistant to drug 1, then drug 2 should be given. Otherwise, it is drug 1 that should be continued, and drug 2 removed. Figure 7.8d illustrates these scenarios. Again, "Strategy 1" corresponds to continuing drug 1, and "Strategy 2" continues with drug 2. Now, it is the potency of drug 1 that is varied (while remaining below the potency of drug 2). By analogy with the previous case, only Strategy 2 is sensitive to changes in H_1, because in this case mutants resistant to drug 2 are only killed in stage I treatment by drug 1, and this has to be done as efficiently as possible. A more interesting observation which confirms the above verbal argument is that "Strategy 1" may or may not be the most efficient. In particular, it is the preferable strategy when drug 2 is significantly more potent than drug 1. In this case, during stage I treatment, the "better" (more specific and more potent) drug (which is drug 2 in our notations), effectively gets rid of the mutants resistant to drug 1. Then the only problem is the mutants resistant to drug 2, which are subsequently killed by a long-term application of drug 1.

We conclude that if drugs with differential activity spectra are used, then the drugs with the broader activity should be continued, and the drug(s) with the narrowest activity can be removed. On the other hand, if the broader activity drug is also considerably more potent, the less potent and narrower drug should remain and the more potent and broader one can be removed.

7.8 Summary

Here we investigated the effect of withdrawing one or more drugs from the combination therapy regimes once the number of cancer cells has declined to lower levels. On the most basic level, the modeling tells us that the number of drugs can be reduced for long-term therapy without significantly reducing the chances of success. This is important in the face of side effects that are likely to become problematic if treatment is continued in the long term. If two drugs have differential activity spectra, and the drug with the broader activity is less potent, then the more specific drug should be removed first. If on the other hand, the broader drug is characterized by a much higher potency, then the broader one should be removed. We further show that the cancer size at which the number of drugs can be reduced does not correlate with the two phases of CML decline. Neither does it depend much on kinetic parameters of CML growth, except for the mutation rates. This is good news because even without any information on most parameters, and using only the data on the cancer size and the mutation rate, it is possible to predict at which stage of treatment the number of drugs can be reduced.

Problems

7.1. Derive formula (7.1) by solving the system for characteristics.

7.2. Research project
Investigate the different methods of calculating the quantity N_{off}, the tumor size at which the number of drugs can be reduced by one without affecting the probability of treatment success. In particular, compare the deterministic method described in Sect. 7.2.2 and the graphical method using the diagram in Fig. 7.5.

7.3. Research project
Find out about the biological mechanisms of cellular quiescence.

7.4. Numerical project
Modify the program obtained for numerical project 4.8 to include a two-phase treatment. Explore how a reduction in the number of drugs influences the probability of treatment success, for different values of t_{**}. Obtain the associated tumor size, N_{off} for which it is safe to reduce the number of drugs. Observe how this size depends on various system parameters.

References

1. Komarova, N.L., Wodarz, D.: Stochastic modeling of cellular colonies with quiescence: an application to drug resistance in cancer. Theor. Popul. Biol. **72**(4), 523–538 (2007)
2. Komarova, N.L., Wodarz, D.: Effect of cellular quiescence on the success of targeted cml therapy. PLoS ONE **2**(10), e990 (2007)
3. Komarova, N.L., Wodarz, D.: Drug resistance in cancer: principles of emergence and prevention. Proc. Natl. Acad. Sci. U.S.A. **102**(27), 9714–9719 (2005)
4. Komarova, N., Wodarz, D.: Combination therapies against chronic myeloid leukemia: short-term versus long-term strategies. Cancer Res. **69**(11), 4904–4910 (2009)

Chapter 8
Cross Resistance: Treatment and Modeling

Abstract We have so far considered combination therapy under the assumption that resistance against one drug does not confer resistance to any of the other drugs in use. While this applies to many of the point mutations that induce resistance against imatinib, dastinib, and nilotinib, one mutation, T315I, confers resistance to all three drugs, i.e., there is complete cross-resistance. This chapter examines the effectiveness of combination therapy in the presence of the cross-resistant T315I mutation. The combination of two drugs is found to increase the probability of treatment success despite this cross-resistance. Combining more than two drugs, however, does not provide further advantages. Hence, according to the model, only the two most effective drugs should be used in combination for the prevention of drug resistance. We also discuss possible combinations involving inhibitors under development that aim to overcome the T315I mutation.

Keywords Keywords Cancer combination treatment · Small molecule inhibitors · CML · Imatinib · Dasatinib · Nilotinib · Cross-resistance · T315I mutation · Mutation rates · Drug dosage · Combinatorial mutation diagram · Drug-resistance networks · Doubly-resistant mutants · Probability of treatment success

8.1 Introduction

The previous chapters have shown how the combination of three or four tyrosine kinase can potentially overcome drug resistance even during the advanced blast crisis stage of CML. It was assumed that mutations that confer resistance against one drug do not confer resistance against any of the other drugs in use. In addition to imatinib, the second-generation inhibitors dasatinib and nilotinib are alternative inhibitors of the BCR-ABL gene product. More than 50 mutations have been identified that confer CML resistance against one of the drugs (in particular imatinib), but not against the others [1]. However, one particular mutation has been identified, called T315I

[2], which confers resistance to all three drugs: imatinib, dasatinib, and nilotinib. This obviously complicates the use of combination therapy. This chapter examines the effectiveness of combining the currently used inhibitors under the assumption that the T315I mutation confers cross-resistance to all drugs, while on the order of 10–100 mutations confer resistance to only one drug in the combination. It also explores the consequences of combining existing drugs with new drugs that can potentially inhibit T315I mutant cells [2].

8.2 The Problem of Cross-Resistance in CML

Much research has recently been devoted to understanding the mechanisms of drug resistance in CML. Drugs in different combinations and different concentrations have been used in in vitro experiments to uncover the principles of resistance [1, 3–7] and to suggest ways to avoid it. In papers [8, 9], a quantitative analysis of mutations has been performed. In in vitro experiments described in these papers, CML cancer cells, Ba/F3 p210$^{bcr-able}$ were exposed to a minimally cytotoxic agent, N-ethyl-N-nitrosourea (ENU), a potent inducer of point mutations. The cells were then cultured in 96-well plates supplemented with graded concentrations of inhibitors. After some time (about 28 days), wells with positive outgrowth were expanded and then sequenced for mutations. Three different inhibitors, imatinib, dasatinib, and nilotinib, were used, in different combinations and solo. Inhibitor concentrations used for the three inhibitors are listed in Table 8.1. The noted concentrations were motivated by the fact that nilotinib is at least 20-fold and dasatinib at least 300-fold more potent than imatinib [8]. After analysis of the total of 768 wells, there were 726 mutations. Of the 30 specific point mutations that had been previously identified in imatinib resistant patients, 25 were recovered in this experiment. In total, 26 point mutations were identified.

In vitro experiments suggest that different concentrations of a drug give rise to different numbers and types of resistant mutants in treating CML. They also show that some mutants are resistant to more than one drug. Figure 8.1 summarizes these fundings. In Fig. 8.1a we show the number of doubly resistant mutants that were discovered in [8] for different combinations of drug doses. The higher the dose, the better cells respond, and the fewer mutations are capable of conferring resistance to the drug. Figure 8.1b shows the numbers of triply resistant mutants.

Table 8.1 Categorization of the doses of each inhibitor, as used in [8]

	Low Dose (nM)	Medium Dose (nM)	High Dose (nM)
Imatinib	2000	4000–8000	16000
Dasatinib	5	10–25	100–500
Nilotinib	50–250	500–1000	2000–5000

(a)

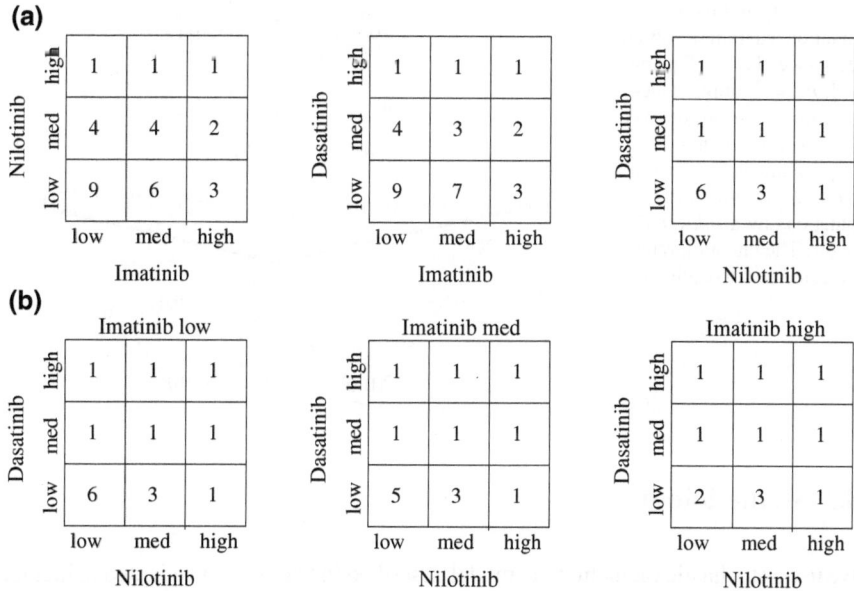

Fig. 8.1 The number of doubly and triply resistant mutants found in the experiments of [8]. (**a**) Doubly resistant mutants. There are three combinations of two drugs, whose doses are marked as "low", "med", and "high". The number of mutants conferring resistance to both drugs is shown in the cells of the tables. (**b**) Triply resistant mutants. The three tables correspond to low, medium, and high doses of imatinib. The doses of nilotinib and dasatinib are indicated on the sides of the tables

In particular, we can see that (unless dasatinib is administered at a low dose) there is only one mutation that confers resistance to all the three drugs. This mutation, termed T315I, has been described in many studies [2, 8, 10–12], and is among the most pressing challenges in the therapy of CML. This mutation is responsible for up to 25 % of all clinically observed resistances in CML patients undergoing imatinib therapy. In addition to nilotinib and dasatinib, it also confers resistance to the newer drug Bosutinib [13]. An important goal is to come up with inhibitors that are effective against T315I mutants [14–21]. One of the most advanced clinical compound currently is danusertib, which exhibits inhibitory activity against the Bcr-Abl tyrosine kinase, including its multidrug-resistant T315I mutant [14]. Along with that, ponatinib is a structurally novel tyrosine kinase inhibitor that potently inhibits the T315I mutation [15]. These and other drugs are currently in different stages of clinical trials.

In this chapter, we examine whether a combination of the three existing drugs (imatinib, dasatinib, and nilotinib) can improve treatment outcome. We also explore the influence of future drugs on treatment outcome.

Fig. 8.2 Combinatorial
mutation diagram for different
resistance classes. Treatment
with $m = 3$ drugs. One-
drug mutations are denoted
by *thin solid arrows*, two-
drug mutations by *dashed
arrows*, and three-drug
mutations by a *thick solid
arrow*. The mutation rates are
indicated next to each arrow

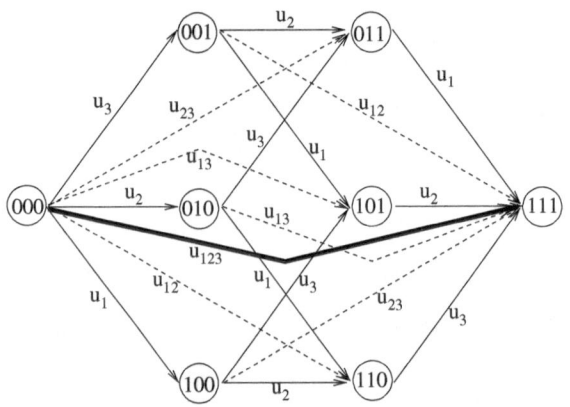

8.3 Methodology

We use a stochastic mathematical model described in Chap. 4, and adapted to include
the presence of cross-resistance, see [22].

Cross-resistance is characterized by the existence of mutants simultaneously resistant to more than one drug. In terms of the combinatorial mutation diagrams, this
manifests itself in "short-cuts"—mutational steps that create resistance to several
drugs at once, see Fig. 8.2. The existing medical knowledge on the types of mutations for CML drugs can be incorporated in the rates of mutations u. One example
is the drugs imatinib, dasatinib, and nilotinib. We assume that k_i mutations give rise
to resistance to drug i without affecting resistance properties with respect to the
other drugs (the index i takes values $1, 2, 3$ for the three drugs). This translates into
$u_i = k_i u$, where u is the basic point mutation rate. Similarly, let us denote by u_{ij}
with $i \neq j$ the rate of mutations leading to cells simultaneousy resistant to drugs i
and j (for three drugs, there are three such rates, u_{12}, u_{13}, and u_{23}). Finally, u_{123} is
the rate of generation of mutants conferring resistance to all the three drugs.

For example, consider Fig. 8.1, and assume that drugs imatinib, nilotinib, and
dasatinib are all administered at a high-dose level; we will refer to these three drugs
by numbers $1, 2$, and 3 respectively. For the experimental conditions reflected in this
figure, we can see that there is only $k_{123} = 1$ mutation giving rise to triple cross-
resistance (Fig. 8.1b). The numbers of mutations conferring resistance to two out of
three drugs can be read off from the tables in Fig. 8.1a. In particular, we have $k_{12} = 3$,
$k_{13} = 3$, and $k_{23} = 1$. These numbers are used to calculate the mutation rates for the
different connections in Fig. 8.2. For example, the arrow connecting types 000 and
110 will have the mutation rate $k_{12}u = 3u$ in this case. With these modifications,
the formalism described in Chap. 4 can be extended in an obvious way to include
cross-resistance.

Different drugs can possess different degrees of cross-resistance. For combinations of three drugs, there are five possibilities, depending on pairwise cross-resistance

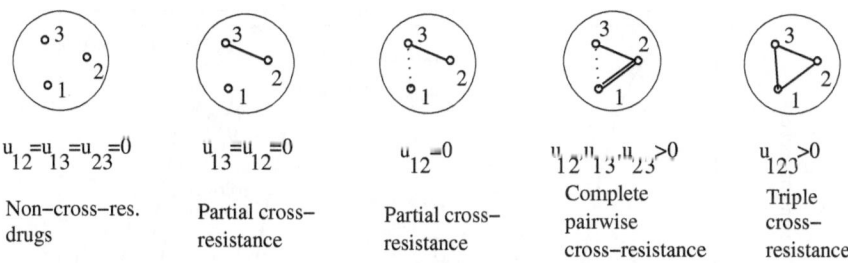

Fig. 8.3 All possible three-drug resistance networks. The number of nodes corresponds to the number of drugs used. Connected nodes correspond to the existence of a cross-resistant mutation. Identical connecting lines indicate that the same mutation confers resistance to all connected drugs. Different (*single, double, dashed*) lines correspond to different mutations

properties of the drugs and the triple-resistance properties. These five cases can be captured by means of five possible three-drug cross-resistance networks, see Fig. 8.3. In the figure, each drug is represented as a node of the network. Each pair of circles is connected if there exists a mutation event that confers resistance against the two drugs. Different lines (single, double, dashed) correspond to different mutations. In Fig. 8.3, the drugs are denoted by the numerals "1", "2", and "3". If a cross-resistance network has no connecting lines (the leftmost diagram in Fig. 8.3) this means that no cross-resistance takes place. If the lines in a cross-resistance network are different, this means that different mutants are resistant against each pair of drugs. If the lines are the same, then triple cross-resistance takes place (the rightmost cross-resistance network in Fig. 8.3). This case is exemplified by the drugs imatinib, dasatinib, and nilotinib. By setting some of the double-resistance and triple-resistance rates to zero, we can model any of the five possible cross-resistance networks by using the diagram of Fig. 8.2.

8.4 Combination Treatment in the Presence of Cross-Resistance

Let us assume that the colony size at start of treatment is between 10^8–10^{13} cells. The death rate of cells in the absence of treatment is between 0 and 70 % of their birth rate. In the presence of drug therapy, susceptible cells have an additional, drug-induced death rate which can be up to ten times greater than the cells' division rate. In the presence of multiple drugs, the susceptible cells are killed at the rate given by the maximum drug-induced rate in the combination. We assume that there are of the order $k_i = k \sim 10$–100 different point mutations that can confer resistance against only one of the drugs. There is also one particular mutation which can confer resistance against all the drugs in use ($k_{123} = 1$).

First, consider the combination of two drugs. For reference, Fig. 8.4 shows the probability of treatment success for one and two drugs assuming the absence of cross-resistance. This is compared to various cross-resistance scenarios. While the

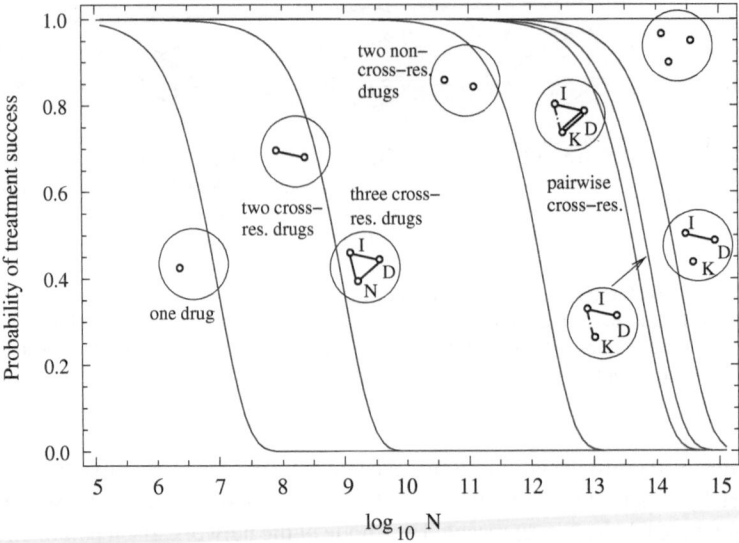

Fig. 8.4 The probability of treatment success is plotted as a function of the colony size, N. Different curves correspond to different combination treatments, with one, two, and three drugs. The cross-resistance networks are presented by using connected and disconnected nodes, see Fig. 8.2. Simulation parameters are as follows: $u = 10^{-9}$, $k = 100$, $M_0 = 100$, $D/L = 0.5$, $H/L = 3$. The symbols "I," "D," "N," and "K" stand for "imatinib," "dasatinib," "nilotinib" and a future drug which can bind to T315 mutants

probability of treatment success is lower in the presence than in the absence of cross-resistance, combining two drugs with cross-resistance clearly improves the probability of treatment success relative to the use of only one drug. The reason is that it is much more likely to acquire a mutation that confers resistance against only one drug than to acquire the doubly resistant mutation. This is because only one specific mutation can lead to cross-resistance, while many mutations can confer resistance against a single drug. Hence, for most mutations, combination therapy will not be challenged by cross-resistance. On a qualitative level, this result does not depend on the kinetic parameters of the model, such as the division and death rates. The advantage of combining two drugs becomes insignificant if the number of mutations that confer resistance to only one drug is very low (Fig. 8.5a), or if the rate at which the doubly resistant mutants are acquired is relatively high. If the number of mutations that confer resistance against only one drug is of the order of 50–100, and if the rate at which resistant mutations are generated is of the order of 10^{-6}–10^{-9}, the model suggests that combining two drugs is advantageous to the patient, even if cross-resistance is possible.

Next, consider a combination of three drugs, i.e., imatinib, dasatinib, and nilotinib. The model shows that combining three drugs will not lead to any further advantage compared to the combination of two drugs (Fig. 8.4). For triple combination therapy to be advantageous, most resistant cells must harbor mutations that render them

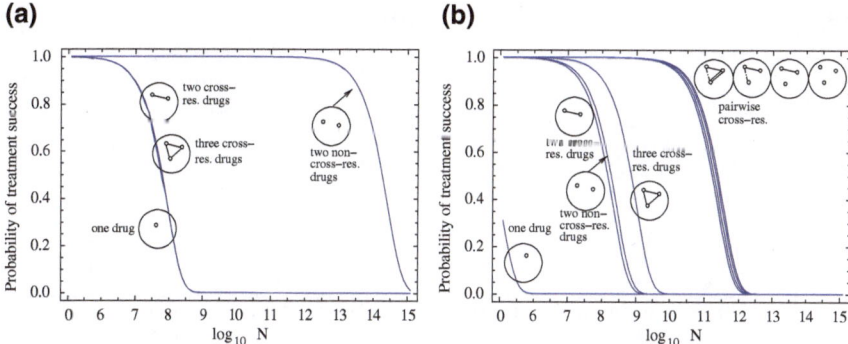

Fig. 8.5 The probability of treatment success is plotted as a function of the colony size, N. (**a**) The number of non-cross-resistant mutations is low ($k = 10$) and the mutation rate for cross-resistance is 10 times higher than in Fig. 8.4. Conclusion: combining more than one cross-resistant drug does not improve the chances of treatment success. (**b**) The number of non-cross-resistant mutations is high, $k = 10^4$. Conclusion: combining three cross-resistant drugs improves the chances of treatment success compared with two cross-resistant or non-cross-resistant drugs (which in turn is better than using only one drug). The rest of the parameters are as in Fig. 8.4

resistant against two of the drugs (but not the third one). Accumulating two separate resistance mutations, however, is a relatively rare event. It is much more likely that a cell acquires the single cross-resistance mutation. Hence, triple combination therapy does not improve the probability of treatment success compared to double combination therapy. Triple combination therapy can only provide an additional advantage if the number of mutations that confer resistance against only one drug is unrealistically high (Fig. 8.5b). Assuming reasonable values for the cellular division, death, and mutation rates, there must be at least $k = 1,000$ mutations that confer resistance against only one drug for triple combination therapy to somewhat improve the chances of treatment success. The improvement becomes significant for $k = 10,000$ or more mutations (Fig. 8.5b). For such high values of k we observe that the effect of cross-resistance is insignificant and treatment failure occurs as a result of mutations that confer resistance to one drug at a time. In this case combining three drugs with triple cross-resistance gives a better outcome than two drugs in the absence of cross-resistance (see Fig. 8.5b). Again, these results do not qualitatively depend on the kinetic parameters of the model. Figure 8.6 demonstrates how the probability of treatment success for one, two and three cross-resistant drug therapies changes as a function of the natural cancer cell death rate (a) and drug-induced death rate (b).

In summary, the analysis shows that in the context of the currently available drugs for CML therapy (imatinib, dasatinib, and nilotinib), combining two of them can provide an advantage over using one drug alone. Experimental data support this notion [8]. However, combining all of these drugs does not improve the chances of treatment success beyond double combination therapy. Hence, the two most effective drugs should be given simultaneously to treat CML.

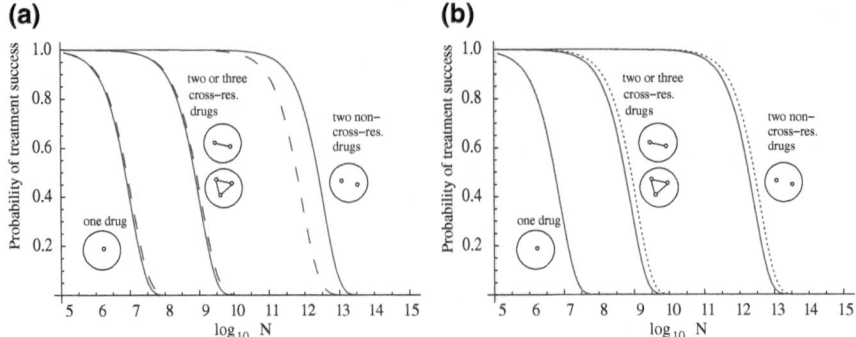

Fig. 8.6   The dependence of treatment success on parameters. (**a**) The role of the natural death-rate of the cancer cells: $D/L = 0$ (*solid lines*) and $D/L = 0.8$ (*dashed lines*); $H/L = 10$. (**b**) The role of the drug-induced death rate: $H/L = 3$ (*solid lines*) and $H/L = 400$ (*dotted lines*); $D/L = 0$. Other parameters are the same as in Fig. 8.4

While these results are robust and do not break down in the context of biologically reasonable values of the cell death rates and drug-induced death rates (Fig. 8.6), the largest degree of uncertainty exists regarding the relationship between the number of tumor cells and the probability of treatment success. As is evident in Fig. 8.4, the probability of treatment success is only high if the number of tumor cells upon start of therapy is no larger than 10^{10}. However, many more tumor cells than this are found in patients when the disease becomes detectable, and especially during more advanced stages. On the other hand, the tumor cell population is heterogeneous, and it is not clear to what extent different cell populations contribute to the disease dynamics [23, 24]. For example, if only the more primitive tumor stem cells and progenitor cells are important determinants of tumor growth, then the effective population size that our model refers to is significantly lower than the overall number of tumor cells detected in a patient.

General cross-resistance networks have relevance for future generation treatment options. If drugs are used that show no cross-resistance with imatinib, dasatinib, and nilotinib, such as VX-680, danusertib, and ponatinib [14–21], would it be advantageous to combine such a drug with two or more drugs that do not inhibit the T315I mutation? Our calculations show that it is advantageous to combine such a future generation drug (call it drug "K") with two drugs that cannot act on the T315I mutation, say, imatinib (I) and dasatinib (D). This is seen for example in Fig. 8.4 when we compare the probability of treatment success for two cross-resistant drugs with that for "pairwise cross-resistant" drugs. We observe that any of the pairwise cross-resistant combinations of three drugs gives a significant improvement. Note that adding a third cross-resistant drug does not improve the chances of successful therapy.

We can examine all possible resistance networks for three-drug therapies, see Figs. 8.4 and 8.6. Let us suppose that there is some cross-resistance between drug K and, say, I. In other words, even though the T315I mutation does not confer resistance to K, there may be a different mutation which makes the cell resistant to both K and I, but not to D. In this situation, treating with I and K is obviously better than just treating with I (this follows from our previous results), but what is interesting, adding D does give an advantage in this case. Adding D will even give an advantage if there is a third mutation which confers resistance to both K and D (but not I). In other words, even though treating with three drugs characterized by triple cross-resistance does not improve the chances of treatment success compared to two such drugs, treating with three pairwise-cross-resistant drugs is advantageous compared to treating with two such drugs.

8.5 Summary

To conclude, we performed quantitative studies of drug cross-resistance in CML treatment. The mathematical framework created here is general enough to be applied to other cancers. However, the model analyzed is specific to CML and the existing drugs imatinib, nilotinib, and dasatinib. We predict that (i) combining two cross-resistant drugs improves the chances of treatment success, (ii) combining more than two drugs that do not inhibit the T315I mutation does not increase the probability of treatment success compared to combinations of two drugs, and (iii) once a drug effective against T315I mutants becomes available, the most effective treatment strategy is to combine that drug with two of the three presently existing drugs. A more detailed model of cross-resistance which involves different drug concentrations can be found in [25].

Problems

8.1. Using Fig. 8.1, find out which of the existing CML drugs possess the highest and the lowest specificity. Given this information, what treatment strategy can be preferable according to the theory presented in Chap. 7 (assuming that the drugs have similar potency, and assuming the absence of cross-resistance)? Which drug should be discontinued after time t_{**}?. Where t_{**} denotes Treatment Stage II

8.2. Draw the combinatorial mutation diagram including cross-resistance, in the case of $m = 2$ drugs (similar to the diagram of Fig. 8.2). For this case:
(a) Write down the Kolmogorov forward equation, similar to that of Sect. 4.3.
(b) Derive the equations for the moments, similar to those of Sect. 4.3.1.
(c) Introduce the probability generating function and write down the equations for the characteristics, similar to those of Sect. 4.3.2.

8.3. Research project
Find out which CML drugs that are effective against T315I mutants are currently available in clinical trials or in the market.

8.4. Numerical project
Modify the program obtained for numerical project 4.8 to include the phenomenon of cross-resistance. Compare treatment success for different values of m, under various assumptions on the mutation network, and different rates of mutations. For $m = 3$, very different results are expected in the presence of a nonzero triple-mutation rate u_{123}.

References

1. Rix, U., Hantschel, O., Dürnberger, G., Rix, L., Planyavsky, M., Fernbach, N., Kaupe, I., Bennett, K., Valent, P., Colinge, J., et al.: Chemical proteomic profiles of the bcr-abl inhibitors imatinib, nilotinib, and dasatinib reveal novel kinase and nonkinase targets. Blood **110**(12), 4055–4063 (2007)
2. O'Hare, T., Eide, C., Deininger, M.: Bcr-abl kinase domain mutations, drug resistance, and the road to a cure for chronic myeloid leukemia. Blood **110**(7), 2242–2249 (2007)
3. Kantarjian, H., Giles, F., Wunderle, L., Bhalla, K., O'Brien, S., Wassmann, B., Tanaka, C., Manley, P., Rae, P., Mietlowski, W., et al.: Nilotinib in imatinib-resistant cml and philadelphia chromosome-positive all. N. Engl. J. Med. **354**(24), 2542–2551 (2006)
4. Deguchi, Y., Kimura, S., Ashihara, E., Niwa, T., Hodohara, K., Fujiyama, Y., Maekawa, T.: Comparison of imatinib, dasatinib, nilotinib and inno-406 in imatinib-resistant cell lines. Leuk. Res. **32**(6), 980–983 (2008)
5. Quintas-Cardama, A., Kantarjian, H., Jones, D., Nicaise, C., O'Brien, S., Giles, F., Talpaz, M., Cortes, J.: Dasatinib (bms-354825) is active in philadelphia chromosome-positive chronic myelogenous leukemia after imatinib and nilotinib (amn107) therapy failure. Blood **109**(2), 497–499 (2007)
6. Vajpai, N., Strauss, A., Fendrich, G., Cowan-Jacob, S., Manley, P., Grzesiek, S., Jahnke, W.: Solution conformations and dynamics of abl kinase-inhibitor complexes determined by nmr substantiate the different binding modes of imatinib/nilotinib and dasatinib. J. Biol. Chem. **283**(26), 18,292–18,302 (2008)
7. Hantschel, O., Rix, U., Superti-Furga, G.: Target spectrum of the bcr-abl inhibitors imatinib, nilotinib and dasatinib. Leuk. Lymphoma **49**(4), 615–619 (2008)
8. Bradeen, H., Eide, C., O'Hare, T., Johnson, K., Willis, S., Lee, F., Druker, B., Deininger, M.: Comparison of imatinib mesylate, dasatinib (bms-354825), and nilotinib (amn107) in an n-ethyl-n-nitrosourea (enu)-based mutagenesis screen: High efficacy of drug combinations. Blood **108**(7), 2332–2338 (2006)
9. O'Hare, T., Eide, C., Tyner, J., Corbin, A., Wong, M., Buchanan, S., Holme, K., Jessen, K., Tang, C., Lewis, H., et al.: Sgx393 inhibits the cml mutant bcr-ablt315i and preempts in vitro resistance when combined with nilotinib or dasatinib. Proc. Nat. Acad. Sci. **105**(14), 5507–5512 (2008)
10. Talpaz, M., Shah, N., Kantarjian, H., Donato, N., Nicoll, J., Paquette, R., Cortes, J., O'Brien, S., Nicaise, C., Bleickardt, E., et al.: Dasatinib in imatinib-resistant philadelphia chromosome-positive leukemias. N. Engl. J. Med. **354**(24), 2531–2541 (2006)
11. Deininger, M.: Optimizing therapy of chronic myeloid leukemia. Exp. Hematol. **35**(4), 144–154 (2007)

12. Weisberg, E., Manley, P., Cowan-Jacob, S., Hochhaus, A., Griffin, J.: Second generation inhibitors of bcr-abl for the treatment of imatinib-resistant chronic myeloid leukaemia. Nat. Rev. Cancer **7**(5), 345–356 (2007)

13. Gontarewicz, A., Brümmendorf, T.: Danusertib (formerly pha-739358)-a novel combined pan-aurora kinases and third generation bcr-abl tyrosine kinase inhibitor. Small Molecules in Oncology, pp. 199–214. Springer, Heidelberg (2010)

14. Winter, G., Rix, U., Carlson, S., Gleixner, K., Grebien, F., Gridling, M., Müller, A., Breitwieser, F., Bilban, M., Colinge, J., et al.: Systems-pharmacology dissection of a drug synergy in imatinib-resistant cml. Nat. Chem. Biol. **8**(11), 905–912 (2012)

15. Pinilla-Ibarz, J., Flinn, I.: The expanding options for front-line treatment in patients with newly diagnosed cml. Crit. Rev. Oncol. Hematol. (2012)

16. Burley, S.: Application of fasttm fragment-based lead discovery and structure-guided design to discovery of small molecule inhibitors of bcr-abl tyrosine kinase active against the t315i imatinib-resistant mutant. In: Proceedings of the American Association for Cancer Research, 2006(1), pp. 1139 (2006)

17. Carter, T., Wodicka, L., Shah, N., Velasco, A., Fabian, M., Treiber, D., Milanov, Z., Atteridge, C., Biggs III, W., Edeen, P., et al.: Inhibition of drug-resistant mutants of abl, kit, and egf receptor kinases. In: Proceedings of the National Academy of Sciences of the United States of America, 102(31), pp. 11,011–11,016 (2005)

18. Melo, J., Chuah, C.: Resistance to imatinib mesylate in chronic myeloid leukaemia. Cancer Lett. **249**(2), 121–132 (2007)

19. Harrington, E., Bebbington, D., Moore, J., Rasmussen, R., Ajose-Adeogun, A., Nakayama, T., Graham, J., Demur, C., Hercend, T., Diu-Hercend, A., et al.: Vx-680, a potent and selective small-molecule inhibitor of the aurora kinases, suppresses tumor growth in vivo. Nat. Med. **10**(3), 262–267 (2004)

20. Young, M., Shah, N., Chao, L., Seeliger, M., Milanov, Z., Treiber, D., Patel, H., Zarrinkar, P., Lockhart, D., et al.: Structure of the kinase domain of an imatinib-resistant abl mutant in complex with the aurora kinase inhibitor vx-680. Cancer Res. **66**(2), 1007–1014 (2006)

21. Duncan, E., Goetz, C., Stein, S., Mayo, K., Skaggs, B., Ziegelbauer, K., Sawyers, C., Baldwin, A.: Iκb kinase β inhibition induces cell death in imatinib-resistant and t315i dasatinib-resistant bcr-abl+ cells. Mol. Cancer Ther. **7**(2), 391–397 (2008)

22. Komarova, N., Wodarz, D.: Combination therapies against chronic myeloid leukemia: short-term versus long-term strategies. Cancer Res. **69**(11), 4904–4910 (2009)

23. Wicha, M., Liu, S., Dontu, G.: Cancer stem cells: an old ideaa paradigm shift. Cancer Res. **66**(4), 1883–1890 (2006)

24. Reya, T., Morrison, S.J., Clarke, M.F., Weissman, I.L.: Stem cells, cancer, and cancer stem cells. Nature **414**(6859), 105–11 (2001)

25. Katouli, A., Komarova, N.: Optimizing combination therapies with existing and future cml drugs. PloS one **5**(8), e12,300 (2010)

Chapter 9
Mathematical Modeling of Cyclic Cancer Treatments

Abstract We have so far discussed combination therapies as a means to overcome drug resistance. While the simultaneous administration of multiple drugs is the most effective way to avoid treatment failure, this might not always be possible, e.g., due to issues related to tolerance and toxicity. Here, we examine cyclic treatment regimes, which involve alternating applications of two (or more) different drugs, given one at a time. We discuss how mathematical methods can help us identify guidelines on optimal cyclic treatment scheduling, with the aim of minimizing resistance generation. We define a condition on the drugs' potencies which allows for a relatively successful application of cyclic therapies. We find that the best strategy is to start with the stronger drug, but use longer cycle durations for the weaker drug. We further investigate the situation where a degree of cross-resistance is present, such that certain mutations cause cells to become resistant to both drugs simultaneously. We show that the general rule (best-drug-first, worst-drug-longer) is unchanged by the presence of cross-resistance. We design a systematic method to test all strategies and come up with the optimal timing and drug order. The role of various constraints on the optimal therapy design, and in particular, suboptimal treatment durations and drug toxicity, is considered.

Keywords Cancer combination treatment · Cyclic treatment · Drug toxicity · Side effects · Best drug first · Worst drug first · Activity spectra · Potency of drugs · Specificity of drugs · Optimization · Objective function · Deterministic and stochastic modeling · Mutually-strong drugs · Treatment protocols

9.1 Introduction

Mathematical modeling of cancer therapy performed by Goldie and Coldman [1–5] and Day [6] in the late 1970s and the 1980s had a widespread impact on the design of new chemotherapy regimens for testing in clinical trials [7]. This was

N. L. Komarova and D. Wodarz, *Targeted Cancer Treatment in Silico*,
Modeling and Simulation in Science, Engineering and Technology,
DOI: 10.1007/978-1-4614-8301-4_9, © Springer Science+Business Media New York 2014

an exciting time in the history of mathematical modeling in medicine. Stochastic models applied to generation of drug resistance and treatment optimization have been used in scheduling chemotherapy treatments. In particular, wet lab oncologists undertook the task of testing a specific, purely theoretical rule proposed by [6]: the "worst drug rule".

The worst drug rule is a general optimization principle which states that if two non-cross-resistant drugs are used intermittently in treatment, then the drug with the weaker killing rate has to be applied first and/or for a longer period of time. This rule has been derived computationally, in the context of non-cross-resistant drugs, by considering 16 different treatment strategies listed below:

$AAAAAAAAAAAA$ $BBBAAAAAAAAA$ $BABABABABABA$ $ABBBABBBABBB$
$AAAAAAAAABBB$ $AAAAAABBBBBB$ $BBBAAABBBAAA$ $BBBABBBABBBA$
$AAABAAABAAAB$ $AAABBBAAABBB$ $BBBBBBAAAAAA$ $BBBBBBBBBBAAA$
$BAAABAAABAAA$ $ABABABABABAB$ $AAABBBBBBBBB$ $BBBBBBBBBBBB$

Here, each letter "A" or "B" denotes a single course of treatment with drug A or B respectively; a course is assumed to last one month. Each treatment strategy consists of the total of 12 courses (which fixes the length of treatment). The parameters were chosen such that the total length of treatment is enough to eradicate all the susceptible mutants. For each of the treatment strategy, the probability of treatment success was calculated for different coefficient values, using the methodology developed by [8]. In particular, the mutation rates, the killing rates, and the proliferation rates were varied. The conclusion of the study was that, with a few exceptions, the most successful strategies are the ones where the "worst" drug (that is, the drug with a weaker killing rate) was applied earlier, or for a longer duration, compared to the better drug. The authors distinguish two variants of the worst drug rule [7]: (a) "use more of the worst drug" and (b) "use the worst drug first".

Cyclic treatment strategies similar to the ones described above, are still used for treatment of CML and other cancers. These strategies involve a sequential application of several (usually, two) different drugs. In this chapter, we present an update on the issue of cyclic treatment optimization, and pose a question similar to that raised by [6]: what is the optimal timing of treatment in the context of cyclic therapy? Our analysis employs a wider variety of methods and is more general and systematic. In particular, instead of testing a subset of 16 treatment strategies, we test all cyclic treatment strategies to find the optimal cycle duration and the drug sequence. In addition, we extend the studies to drugs with cross-resistance.

9.2 Methodology

In this section we will adapt the general framework developed in Chap. 4 to describe a cyclic two-drug treatment protocol. Cyclic treatment considered here consists of a sequence of $2\mathcal{N}$ alternating applications of drugs 1 and 2. As described above, before treatment starts (that is, for $0 < t < t_*$), we have $d_s = D_s$; normally, $D_s < L_s$, that is, the cancer is assumed to grow stochastically before treatment. At time t_*, the first drug is applied for a length of time, Δt_1. During this time, we have

$$d_{00} = D_{00} + H_1, \quad d_{10} = D_{10}, \quad d_{01} = D_{01} + H_1, \quad d_{11} = D_{11}, \qquad (9.1)$$

where H_1 is the drug-induced death rate for drug one; see Fig. 9.1b, on the left. After time $t_* + \Delta t_1$, drug 2 is applied for the duration Δt_2, resulting in

$$d_{00} = D_{00} + H_2, \quad d_{10} = D_{10} + H_2, \quad d_{01} = D_{01}, \quad d_{11} = D_{11}, \qquad (9.2)$$

for $t_* + \Delta t_1 < t < t_* + \Delta t_1 + \Delta t_2$ (Fig. 9.1b on the right). Here, H_2 denotes the drug-induced death rate of drug two. After that, treatment is again switched to drug 1 for duration Δt_1, and so on, for a total time duration T_{treat}, with a total of $2\mathcal{N}$ cycles.

We assume that the division rates, L_s, and the death rates, D_s, of cells are time-independent. We will further assume that $L_s = L$ and $D_s = D$, and denote

$$\gamma = L - D.$$

As before, the time of the start of treatment, t_*, can be related to the initial colony size, N, by the deterministic relationship, $N = M_0 e^{(L-D)t_*}$.

9.2.1 Deterministic Modeling of Cyclic Treatments

Here we create a simplified, deterministic framework for studying the effects of cyclic drug treatments. We start by writing down the equations for the expected numbers of different types of resistant mutants, which we call x_s:

$$\dot{x}_{00} = (L(1 - u_1 - u_2 - u_{12}) - D - H_{00})x_{00}, \qquad (9.3)$$
$$\dot{x}_{01} = (L(1 - u_1 - u_{12}) - D - H_{01})x_{01} + Lu_2 x_{00}, \qquad (9.4)$$
$$\dot{x}_{10} = (L(1 - u_2 - u_{12}) - D - H_{10})x_{10} + Lu_1 x_{00}, \qquad (9.5)$$
$$\dot{x}_{11} = (L - D)x_{11} + L[(u_2 + u_{12})x_{10} + (u_1 + u_{12})x_{01} + u_{12}x_{00}], \qquad (9.6)$$

where L and D are the division and natural death rates of the cells, and H_s is the drug-induced death rate for resistance type s; the coefficients H_s change depending on the treatment phase.

We are interested in the production of fully resistant mutants, because (for long-term treatments) such mutants are the reason for treatment failure. These mutants are produced by mutations of partially resistant mutants, and also, in the presence of cross-resistance, by mutations of fully susceptible mutants. This happens both before the treatment starts, and after it starts. The former process is treatment-independent.

Therefore, in order to evaluate the effectiveness of different treatment protocols, it is sufficient to only consider the latter process. The production of fully resistant mutants is described by the second term on the right-hand side of Eq. (9.6). The total amount of doubly-resistant mutants produced after treatment starts is given by

$$F = \int_{t_*}^{T_{\text{treat}}} L[(u_2 + u_{12})x_{10} + (u_1 + u_{12})x_{01} + u_{12}x_{00}]\, dt,$$

where t_* is the time point where therapy is first applied, and T_{treat} is the total treatment duration.

9.2.1.1 Production of Fully Resistant Mutants from Partially-Resistant Mutants

Let us find the functions $x_{10}(t)$ and $x_{01}(t)$ by solving Eqs. (9.4) and (9.5) for $t > t_*$. In our calculations, we will only consider the highest order contributions in terms of the mutation rates.

Let us denote by $A(u_1)$ and $A(u_2)$ the expected number of mutants of types (10) and (01) at the start of treatment. These quantities can be calculated from equations similar to system (9.3–9.6), formulated for the pre-treatment period. In the limit of small mutation rates, we have

$$A(u) \approx N \log Nu,$$

where $N = M_0 e^{(l-d)t_*}$ is the number of cells at the beginning of treatment. Suppose that at the start of treatment, we have $A(u_2)$ mutants of type (01) and $A(u_1)$ mutants of type (10). Starting at $t = t_*$, we treat with drug 1 for time duration Δt_1. This is followed immediately by treatment with drug 2 for time duration Δt_2, then again by treatment with drug 1 for time duration Δt_1, and so on, for an infinite number of cycles.

The average numbers of partially resistant mutants, $x_{10}(t)$ and $x_{10}(t)$, can be found as follows. We define quantities $x_{10}^{(i)}$ and $x_{01}^{(i)}$ for $i = 1, 2, \ldots$ by means of the differential equations:

$$\dot{x}_{01}^{(0)} = (\gamma - H_1)x_{01}, \quad t_* < t < t_* + \Delta t_1, \quad x_{01}(t_*) = A(u_2);$$

$$\dot{x}_{10}^{(0)} = \gamma x_{10}, \quad t_* < t < t_* + \Delta t_1, \quad x_{10}(t_*) = A(u_1);$$

$$\dot{x}_{01}^{(1)} = \gamma x_{01}, \quad t_* + \Delta t_1 < t < t_* + \Delta t_1 + \Delta t_2, \quad x_{01}(t_*) = x_{01}^{(0)}(t_* + \Delta t_1);$$

$$\dot{x}_{10}^{(1)} = (\gamma - H_2)x_{10}, \quad t_* + \Delta t_1 < t < t_* + \Delta t_1 + \Delta t_2, \quad x_{10}(t_*) = x_{10}^{(0)}(t_* + \Delta t_1);$$

$$\ldots$$

To write down the equations for the general cycle i, it is convenient to define the time intervals

$$U_i = \begin{cases} [t_* + \frac{i}{2}(\Delta t_1 + \Delta t_2), t_* + \frac{i}{2}(\Delta t_1 + \Delta t_2) + \Delta t_1), & i \text{ is even}, \\ [t_* + \frac{i-1}{2}(\Delta t_1 + \Delta t_2) + \Delta t_1, t_* + \frac{i+1}{2}(\Delta t_1 + \Delta t_2)), & i \text{ is odd}. \end{cases}$$

Each interval equals a single cycle of a drug; even values of i correspond to cycles of drug 1 and odd values of i to cycles of drug 2. More precisely, interval U_i corresponds to the $(i/2+1)$th application of drug 1 for even values of i, and to the $((i-1)/2+1)$th application of drug 2 if i is odd. The dynamics of cell types are described by an initial value problem within each interval, U_i. We have, if i is even:

$$\dot{x}_{01}^{(i)} = (\gamma - H_1)x_{01}, \quad \dot{x}_{10}^{(i)} = \gamma x_{10}, \quad t \in U_i, \tag{9.7}$$

$$x_{01}(t_* + \frac{i}{2}(\Delta t_1 + \Delta t_2)) = x_{01}^{(i-1)}(t_* + \frac{i}{2}(\Delta t_1 + \Delta t_2)),$$

$$x_{10}(t_* + \frac{i}{2}(\Delta t_1 + \Delta t_2)) = x_{10}^{(i-1)}(t_* + \frac{i}{2}(\Delta t_1 + \Delta t_2));$$

and if i is odd:

$$\dot{x}_{01}^{(i)} = (\gamma)x_{01}, \quad \dot{x}_{10}^{(i)} = (\gamma - H_2)x_{10}, \quad t \in U_i, \tag{9.8}$$

$$x_{01}(t_* + \frac{i-1}{2}(\Delta t_1 + \Delta t_2) + \Delta t_1) = x_{01}^{(i-1)}(t_* + \frac{i-1}{2}(\Delta t_1 + \Delta t_2) + \Delta t_1),$$

$$x_{10}(t_* + \frac{i-1}{2}(\Delta t_1 + \Delta t_2) + \Delta t_1) = x_{10}^{(i-1)}(t_* + \frac{i-1}{2}(\Delta t_1 + \Delta t_2) + \Delta t_1).$$

For any given value of $t > t_*$, we can find i such that $t \in U_i$. Then $x_{10}(t) = x_{10}^{(i)}(t)$, and $x_{01}(t) = x_{01}^{(i)}(t)$. The quantity of interest is the total amount of the fully resistant types produced, which, after $2\mathcal{N}$ cycles, is given by:

$$F_{1\to 2}(\mathcal{N}) = L \sum_{i=0}^{2\mathcal{N}} \int_{t \in U_i} [(u_1 + u_{12})x_{01}^{(i)}(t) + (u_2 + u_{12})x_{10}^{(i)}(t)]\, dt,$$

where the subscript $1 \rightarrow 2$ indicates that the mechanism of resistance generation considered here is a mutation of cells resistant to only one drug, into cells resistant to two drugs. Solving Eqs. (9.7) and (9.8) and integrating, we obtain, $F_{1\to 2}(\mathcal{N})/L =$

$$\frac{A(u_2)(u_1 + u_{12})(1 - e^{\mathcal{N}[(\gamma - H_1)\Delta t_1 + \gamma \Delta t_2]})}{\gamma} \left(\frac{(e^{\gamma \Delta t_1} - e^{H_1 \Delta t_1})H_1}{(e^{H_1 \Delta t_1} - e^{\gamma(\Delta t_1 + \Delta t_2)})(\gamma - H_1)} - 1 \right)$$

$$+ \frac{A(u_1)(u_2 + u_{12})(1 - e^{\mathcal{N}[\gamma \Delta t_1 + (\gamma - H_2)\Delta t_2]})}{H_2 - \gamma} \left(\frac{e^{H_2 \Delta t_2}H_2(e^{\gamma \Delta t_1} - 1)}{\gamma(e^{H_2 \Delta t_2} - e^{\gamma(\Delta t_1 + \Delta t_2)})} + 1 \right).$$

$$\tag{9.9}$$

Let us consider the limit of long treatments, $\mathcal{N} \to \infty$. For convergence, we need to assume that $(\gamma - H_1)\Delta t_1 + \gamma \Delta t_2 < 0$, and $\gamma \Delta t_1 + (\gamma - H_2)\Delta t_2 < 0$ (see [9] for an explanation of these assumptions). Then we have

$$
\begin{aligned}
F_{1\to 2}/L = {} & \frac{A(u_2)(u_1 + u_{12})}{\gamma} \left(\frac{(e^{\gamma \Delta t_1} - e^{H_1 \Delta t_1})H_1}{(e^{H_1 \Delta t_1} - e^{\gamma(\Delta t_1 + \Delta t_2)})(\gamma - H_1)} - 1 \right) \\
& + \frac{A(u_1)(u_2 + u_{12})}{H_2 - \gamma} \left(\frac{e^{H_2 \Delta t_2} H_2(e^{\gamma \Delta t_1} - 1)}{\gamma(e^{H_2 \Delta t_2} - e^{\gamma(\Delta t_1 + \Delta t_2)})} + 1 \right).
\end{aligned} \tag{9.10}
$$

9.2.1.2 Production of Fully Resistant Mutants from Fully Susceptible Cells

Let us next consider the production of cells of type x_{11} directly from the cells of type x_{00} by means of mutations with the rate u_{12}. The equations for x_{00} are $\dot{x}_{00} = (\gamma - H_1)x_{00}$ and $\dot{x}_{00} = (\gamma - H_2)x_{00}$ for odd and even cycles, respectively. Solving these equations and integrating, we can calculate the additional contribution to the function F coming from cross-resistance. The result in the limit of long treatments is:

$$
F_{0\to 2}/L = u_{12} \int_0^\infty x_{00}(t)\, dt = \tag{9.11}
$$

$$
u_{12} N \frac{e^{H_1 \Delta t_1 + H_2 \Delta t_2}(\gamma - H_2) - e^{\gamma(\Delta t - 1 + \Delta t_2)}(\gamma - H_1) + e^{\gamma \Delta t_1 + H_2 \Delta t_2}(H_2 - H_1)}{(e^{\gamma(\Delta t_1 + \Delta t_2)} - e^{H_1 \Delta t_1 + H_2 \Delta t_2})(\gamma - H_1)(\gamma - H_2)},
$$

where N is the number of cells at the beginning of treatment; as before, we assume that the mutation rates are small.

9.2.1.3 Optimization of Treatment Strategies

The deterministic method of finding the optimal treatment strategy is based on the minimization of the objective function,

$$
F = F_{1\to 2} + F_{0\to 2}, \tag{9.12}
$$

with respect to parameters Δt_1 and Δt_2. In the absence of cross-resistance, fully resistant mutants are produced only by partially resistant mutants, and we have $F = F_{1\to 2}$ with $u_{12} = 0$. In the presence of cross-resistance, both mechanisms of fully resistant mutant production are in place.

The objective function F is correlated with the probability of treatment failure. Its minimum, as a function of Δt_1 and Δt_2, optimizes the treatment protocol. Finding this minimum is a much easier task compared to that of the stochastic method, where we need to maximize the probability of treatment success, which can only be calculated numerically (e.g., by means of the Runge-Kutta method of solving ordinary differential equations for the characteristics).

9.2.1.4 Method Applicability and Mutually Strong Drugs

The deterministic method described here is very intuitive and easy to implement compared to the stochastic method. While it does not work well for short treatments (see also [9]), in the limit of long treatments it describes reality remarkably well.

To find the conditions of applicability of the deterministic method, let us solve linear differential Eqs. (9.7) and (9.8) to find the dynamics of partially resistant mutants under treatment conditions. After $2\mathcal{N}$ cycles, that is, for $t_{\mathcal{N}} = \mathcal{N}(\Delta t_1 + \Delta t_2)$, we have

$$x_{01}(t_{\mathcal{N}}) = A(u_2)e^{\mathcal{N}[(\gamma - H_1)\Delta t_1 + \gamma \Delta t_2]}, \quad x_{10}(t_{\mathcal{N}}) = A(u_1)e^{\mathcal{N}[\gamma \Delta t_1 + (\gamma - H_2)\Delta t_2]}.$$
(9.13)

Therefore, in order for the treatment in the absence of doubly resistant mutants to work, we need to require that

$$H_1, H_2 > \gamma \quad \text{and} \quad \frac{\gamma}{H_2 - \gamma} < \frac{\Delta t_2}{\Delta t_1} < \frac{H_1 - \gamma}{\gamma}.$$
(9.14)

If conditions (9.14) are satisfied, functions $x_{10}(t)$ and $x_{01}(t)$ will on average decay. Conditions (9.14) are equivalent to condition (9.15), the "mutual strength" discussed below.

9.2.2 Stochastic Modeling of Cyclic Treatments

The stochastis theory developed earlier applies to cyclic treatments without change. The solution $\xi_{00}(t)$ is obtained from system (4.18–4.21). It consists of coupled non-linear equations of the Riccati type, and will be solved numerically. This system is characterized by piecewise-constant coefficients, and as a consequence, the time-reversal procedure involved in the solution, requires the following approach. Suppose for simplicity that the total duration of treatment equals $\mathcal{N}(\Delta t_1 + \Delta t_2)$, $\mathcal{N} \in I$, that is, it includes an integer and equal number of cycles of each drug. We start by solving system (4.18–4.21) with constant coefficients given by Eq. (9.2) and with initial conditions (4.22), and find the solution corresponding to the time Δt_2. Then we plug these values as initial conditions for system (4.18–4.21) with coefficients (9.1) and solve the equations for time duration Δt_1. Next, we use the resulting functions as the initial conditions and repeat the process for the total of $2\mathcal{N}$ alternating cycles. This procedure corresponds to the ("reversed") time-variable changing from the end of treatment (physical time $t_* + \mathcal{N}(\Delta t_1 + \Delta t_2)$) back to time t_*. Finally, we plug the obtained values in the same system with coefficients corresponding to the pre-treatment conditions ($D_s = d_s$ for all s) and find the solution at $t = t_*$. The obtained function ξ_{00} is used in formula (4.10).

9.3 Mutually Strong Drugs and Treatment Optimization

9.3.1 Modeling Cyclic Treatment Strategies: Drug Potencies and Activity Spectra

Cyclic drug treatments are assumed to proceed as follows, see Fig. 9.1a. Treatment starts at time t_*. Drug 1 is applied for a time duration of Δt_1. Then the drug is discontinued and replaced by drug 2. After time duration Δt_2, drug 2 is in turn replaced by drug 1. The total treatment duration is denoted by T_{treat} and consists of $2\mathcal{N}$ cycles of treatment (here the word "cycle" refers to a one-drug treatment with drug 1 or 2).

Mathematically, each treatment protocol corresponds to specific values of the death rates, D_s, at different moments of time:

$$d_s = D_s + H_s(t),$$

where as before, the coefficients D_s are natural death rates of the cancer cells, and $H_s(t)$ are the drug-induced cell death rates. The functions $H_s(t)$ depend on the

(a)

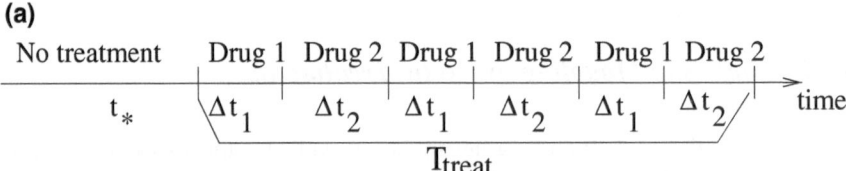

(b)

Fig. 9.1 Cyclic two-drug treatments. (**a**) A cyclic treatment protocol. (**b**) The mutation diagrams and drug-induced death rates for treatments with drug 1 and drug 2. The resistance types are represented by circles with binary indices; the drug-induced death rates are marked next to the circles; the mutation rates are indicated next to the arrows

particular treatment strategy used. As different drugs are applied, the "strength" of each drug, which depends on the concentration of the drug in the patient's blood, changes as some smooth function of time. The exact shape of these functions, and therefore, the shape of $H_s(t)$, depends not only on the treatment strategy (that is, whether drugs are applied in combination, or cyclically), but also on the way the drugs are administered, and on how quickly they are absorbed. For example, it can be assumed that $H_s(t)$ for a susceptible class reaches a maximum sometime after the drug is taken, and decays until the next administration of the drug. However, in this chapter we simplify this picture by assuming that the functions $H_s(t)$ are piecewise constant. They are assumed to have a constant nonzero value for all the susceptible classes as long as the patient is treated with a given drug, and they become zero after the drug is discontinued. For the effects of pharmacokinetics on the dynamics of treatment see [10].

We will discuss two characteristics of drugs, their potency and their activity spectra (see also Chap. 7). The conventional measure of *drug potency* used *in vitro* is the drug concentration needed to achieve a certain log kill. From the biochemical point of view this is related to the affinity, or on-rates, of the inhibitors. The mathematical definition uses the following way to quantify the potency. For the purposes of this study we assume that one drug is more potent than another if, under the given dosages, it results in a higher drug-induced death rate. The latter is measured as the (exponential) rate of the decay of a colony of target cells exposed to the given dose of the drug. Therefore, in what follows we will assume that higher potencies are correlated with higher values of H_s for susceptible mutant classes.

The *activity spectra* of the drugs are described as follows. Drugs with broader activity spectra are characterized by a smaller number of mutations which confer resistance to such drugs. In other words, these drugs are active against a larger number of mutants (and fail against a smaller number of mutants). On the other hand, the more specific, or narrow, drugs fail against a larger number of mutants. In this context we consider the values of cells' mutation rates, with which various types of mutants are produced. If mutants resistant to drug 1 are produced with rate u_1, and mutants resistant to drug 2 are produced with rate u_2, then the inequality $u_1 < u_2$ means that drug 1 has a broader activity spectrum. That is, in our model drugs with a narrower activity spectrum correspond to larger mutation rates associated with the generation of corresponding resistant mutants.

9.3.2 A Prerequisite of Success: Mutually Strong Drugs

We are interested in maximizing the chances of treatment success, that is, in finding a protocol which is characterized by the highest probability of cancer cell eradication in a patient, while remaining within tolerance bounds in terms of its side effects. In particular, we are interested in the production of fully resistant mutants, because such mutants can become the reason for treatment failure.

The mutation diagrams corresponding to the two different cycles (treatment with drug 1 and treatment with drug 2) are shown in Fig. 9.1b. Depending on which drug is applied, different mutants are experiencing different drug-induced death rates. For example, during the cycle of the first drug application, mutants of type (10) (that is, mutants resistant to drug 1 and susceptible to drug 2) are growing, and mutants of type (01) are killed at rate H_2. Similarly, during during the cycle of the second drug application, mutants of type (10) are killed at rate H_2, and mutants of type (01) are growing.

In Fig. 9.2 we plot the population sizes of the two partially resistant colonies, $x_{10}(t)$ (dashed lines) and $x_{01}(t)$ (solid lines), in the course of treatment, for some fixed values of the parameters. In Fig. 9.2a, treatment starts with drug 1, and we can see that during the first cycle the colony x_{01} which is susceptible to this drug, decays exponentially, while the colony x_{10}, which is resistant to this drug, grows. In the second cycle, when drug 2 is applied, colony x_{01} grows and colony x_{10} decays. In Fig. 9.2b we present the scenario where the order of the drugs is switched (while all the parameters are kept the same). We can see immediately that the dynamics of mutant decay is sensitive to the order in which the two drugs are applied.

Before we study the effect of choosing different strategies, we note that not every cyclic strategy can be successful, even in the absence of fully resistant mutants. It turns out (see Methodology for details) that in order to achieve treatment success, a certain minimal condition on the effectiveness of the two drugs has to be satisfied. This condition is given by

$$\frac{1}{h_1} + \frac{1}{h_2} < \frac{1}{\gamma}, \tag{9.15}$$

and can be viewed as a requirement for the two drugs to be sufficiently strong (compared to the colony growth-rate γ) such that they can eliminate a population of partially resistant mutants. We will refer to drugs that satisfy condition (9.15) as *mutually strong* drugs.

Before we proceed to optimize cyclic treatment protocols, we note an important implications of our model. Combining two drugs instead of using a cyclic treatment protocol will always correspond to a larger probability of colony elimination. Figure 9.2c shows the dynamics of the two partially susceptible colonies under the cyclic treatment protocols of Figs. 9.2a and b together with the dynamics corresponding to the combination treatment where drugs 1 and 2 are applied simultaneously and continuously (dotted lines). The drug-induced death rates of all types are always smaller (or at least not larger) under combination treatment, thus eliminating the two colonies at a faster rate. Again, using two drugs simultaneously may not be feasible because of toxicity factors.

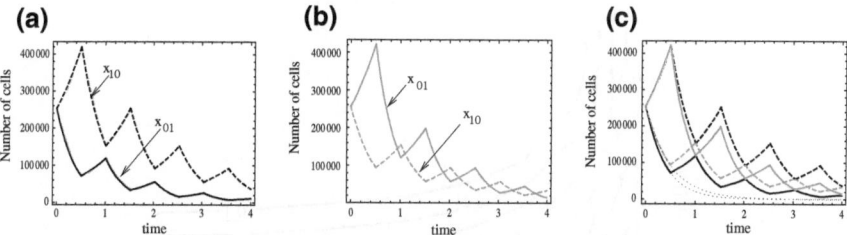

Fig. 9.2 Cyclic two-drug treatments. The deterministic dynamics of the populations of partially resistant mutants, x_{01} (*solid lines*) and x_{10} (*dashed lines*). (**a**) Worst drug first (*black lines*), (**b**) best drug first (*gray lines*), (**c**) lines from panels (**a**) and (**b**) are plotted together; also, the thin dotted lines present the same populations under a combination treatment. The parameters are $H_1 = 3$, $H_2 = 3.5$, $\gamma = 1$, the total number of cells at the start of treatment is $N = 10^{11}$, $u = 10^{-7}$, the numbers of partially resistant mutants at start of treatment are $N_{01} = N_{10} = N \log Nu$, and $\Delta t_1 = \Delta t_2 = \log N/\gamma/50 \approx 0.5$. Neither of the two treatment strategies is optimal

9.4 Analysis of Drug Treatments with Mutually Strong Drugs

Let us fix a certain treatment time, T_{treat}, and vary the number of cycles, $2\mathcal{N}$, used in the protocol. First we consider drugs of different potencies. We would like to determine the following features of the optimal protocol:

(i) What drug should be used first: best drug first (BDF) or worst drug first (WDF)?
(ii) What is the optimal cycle duration ratio for the two drugs?
(iii) What number of cycles, $2\mathcal{N}$, should be implemented within the allocated treatment time?

With a fixed total treatment duration, T_{treat}, let us vary (i) the order of the drugs, (ii) the ratio of the cycle lengths of the two drugs, and (iii) the variable $\mathcal{N}$, which together with (ii) defines the absolute duration of each cycle. For each set of values of the above variables, we will calculate the probability of treatment success, and then find which combination gives the highest such probability.

In Fig. 9.3 we consider two drugs: drug A has potency $H = 5\gamma$ and drug B has potency $H = 3\gamma$; the two drugs have the same activity spectra. The contour plots show the levels of the probability of treatment success calculated by using the stochastic methodology. The horizontal axis represents different values of the number of cycles, and the vertical axis—the ratio of the two cycle lengths. Solid contours correspond to BDF strategies, and dashed contours—to WDF strategies. The numbers next to the contours indicate the probability of treatment success along each contour. For the BDF strategies, the optimal parameter combination corresponds to $\mathcal{N} = 2$ and $\Delta t_B/\Delta t_A \approx 2$, and yields the probability of treatment success around 0.6. For the WDF strategies, the optimal parameter combination corresponds to $\mathcal{N} = 3$ and $\Delta t_B/\Delta t_A \approx 1.5$, and yields the probability of treatment success around 0.4. We can see that treating with BDF has a significant advantage compared to treating with WDF. This is the opposite of the "BDF" rule, but it is consistent with the "use more

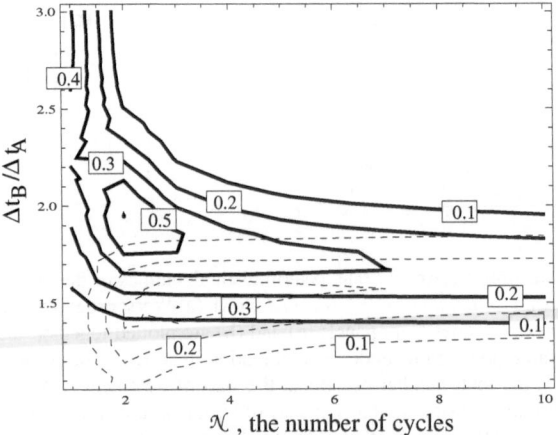

Fig. 9.3 Drugs of different potencies in protocols with BDF and WDF. The horizontal axis is the number of cycles (for a fixed treatment time), and the vertical axis is the ratio of the two drug cycle lengths. For each pair of these parameters, the probability of treatment success is calculated and plotted in the form of a countour plot, with the probability values given by the numbers next to the contours. Solid lines correspond to BDF stratégies, and dashed lines to WDF strategies. Drug A has $H = 5$, drug B has $H = 3$, $T_{treat} = 25$, $L = 1$, $D = 0$, $u = 10^{-8}$, $N = 10^{13}$

of the worst drug" rule, because the cycle duration for the worse drug in the optimal strategy is twice as long as that for the better drug.

Figure 9.4 demonstrates the effects of decreasing the total treatment time. As T_{treat} increases, the probability of treatment success increases (Fig. 9.4a). While for

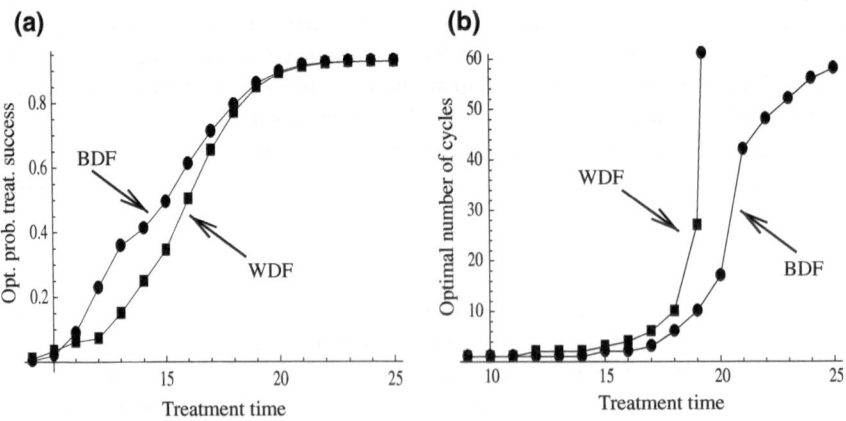

Fig. 9.4 Drugs of different potencies in protocols of different durations. Drug A has $H = 5$, drug B has $H = 3$. (**a**) The optimal probability of treatment success as a function of treatment time. (**b**) The optimal number of cycles as a function of treatment time. The graphs are presented for the WDF and BDF strategies. The other parameters are $L = 1$, $D = 0$, $u = 10^{-8}$, $N = 10^{13}$

any treatment duration, the BDF strategy is always advantageous, the difference between the optimal BDF and WDF treatment protocols becomes smaller for very long treatments. The optimal number of cycles (Fig. 9.4b) increases as the treatment time increases. The optimal BDF protocol usually requires fewer cycles than the optimal WDF protocol. Very short treatment times require the optimal protocol to have only one cycle of drug application. For longer treatments, we notice that the optimal WDF treatment requires an infinite number of cycles, that is, the drugs have to be alternated as frequently as possible, which means that the optimal outcome of the WDF treatment cannot be acieved in practice. A theoretical confirmation of this result was obtained in [9] by using the deterministic methodology which works very well for long treatment times.

Next we will assume that the two drugs have an equal potency and only differ by their activity spectra. As has been mentioned previously, a difference in the activity spectra of two drugs manifests itself in the different rates with which mutants resistant to each drug are produced. In Fig. 9.5, we consider two drugs, drug A with a lower activity spectrum (that is, it is active against fewer mutants), and B with a wider activity spectrum (that is, it is active against a larger number of mutants). Mutants resistant to drug A are produced with rate $u = 10^{-5}$, and mutants resistant to drug B are produced with rate $u = 10^{-9}$. In Fig. 9.5a, we fix the treatment time to be approximately 16.49 times units, which for the parameter values chosen is comparable with the time it takes on average to eliminate the colony of susceptible cells with one of the drugs (approximately 12.66 units). The contour plots of the probability of treatment success obtained by the stochastic method are presented for the two cases: drug A first (solid lines) and drug B first (dashed lines). We observe that the optimal probabilities of treatment success in the two scenarios differ significantly. If we start with drug B (the dashed contours), the optimal strategy is to use two cycles, and

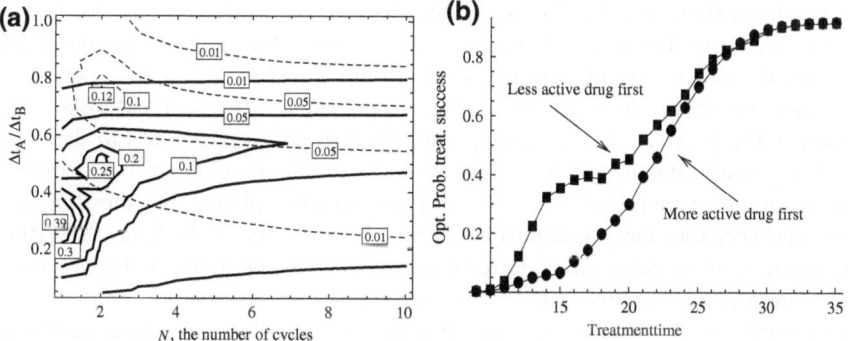

Fig. 9.5 Drugs of different activity spectra in protocols of different duration. Drug A has $u = 10^{-5}$, drug B has $u = 10^{-9}$. (**a**) The contour plots of the probability of treatment success for treatment time is $T_{\text{treat}} = 16.46$; solid lines correspond to $u_1 > u_2$ (drug A first), and dashed lines to $u_1 < u_2$ (drug B first). (**b**) The optimal probability of treatment success as a function of treatment time. The other parameters are $L = 1$, $D = 0$, $H_1 = H_2 = 3$, $N = 10^{11}$

the corresponding success probability is about 0.12. If, on the other hand, we start treatment with drug A (the solid contours), then the best strategy is to use only one cycle, and the corresponding success probability is about 0.39, which is significantly higher. In both cases, the drug with the lower mutation rate must be used for longer.

In Fig. 9.5b, we show that for most finite treatments protocols, it is advantageous to start with the drug characterized by a higher mutation rate, but use the other (more active) drug for longer cycle durations. The difference between the two types of protocols decreases as the treatment length increases, and for very long treatments, the two lines in Fig. 9.5b cross over. In the limit of long treatments, it becomes slightly advantageous to treat with the broader drug first [9].

9.5 Protocols Involving Drugs of High Toxicity

So far we assumed that the objective for treatment protocol optimization is to maximize the number of cancer cells killed. This is a reasonable strategy given that the drugs' side effects are relatively mild. The question of optimization has to be approached differently for treatments with drugs of significant toxicity. In this context, the optimization problem has two objectives: (1) maximize the probability of cancer eradication and (2) minimize the side effects of the treatment.

The framework developed in this chapter is applicable in such situations, but the rules (such as the BDF rule) derived without considerations of toxicity, may not hold. In what follows we describe how to apply our framework to drugs of high toxicity.

By optimizing the probability of cancer eradication, we can produce contour plots of probability of treatment, such as the ones in Fig. 9.3. To demonstrate an alternative set of tools, however, in this section we chose to plot the objective function (see Methodology) which is basically a deterministic tool for figuring out the effectiveness of treatment strategies. The lower the objective function value, the more effective the treatment is. The levels of the objective functions for the BDF and the WDF treatment strategies are plotted in Fig. 9.6. In this figure, we consider a particular example where two drugs are used: drug A has potency $H_A = 5$ and drug 2 has potency $H_B = 3$. Treatment starting with the best drug corresponds to the solid contours, and treatment with the WDF—to dashed contours. Without considerations of toxicity, we simply find the minima of the functions plotted. For the BDF and the WDF scenarios these minima are indicated by black dots in the figure. The BDF treatment (solid contours) has a lower minimum, and the optimum strategy requires using drug B for duration $\Delta t_B \approx 0.55$ and drug A for duration $\Delta t_A \approx 0.40$. The percentages of time when drugs A and B are administered are given by 42 and 58 % respectively.

Before we impose further optimization restrictions coming from the drug toxicity, let us use Fig. 9.6 to compare the two types of treatments (BDF and WDF) for each point $(\Delta t_A, \Delta t_B)$. The thick black line near the top of the contour plot separates the region where the BDF protocols are better (below the line), from the region where WDF protocols are better (above the line). The line corresponds to the points where

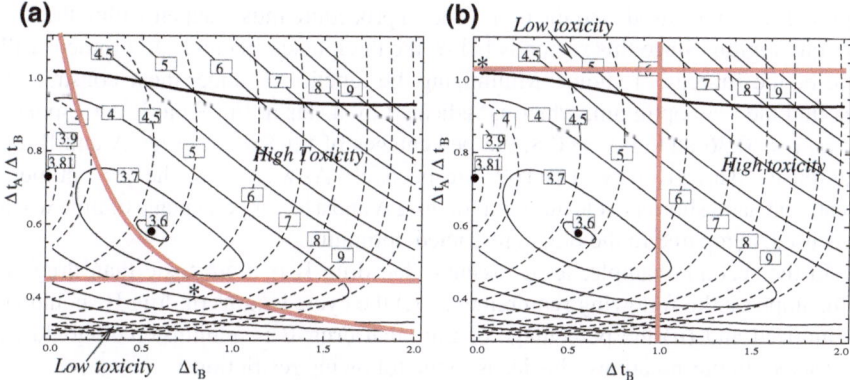

Fig. 9.6 Contour plots of the objective function for the BDF (*solid lines*) and the WDF (*dashed lines*) strategies, for drugs of high toxicity. The lower the value of the objective function, the more effective is the treatment. The axes are Δt_B (the cycle duration of the worse drug) and $\Delta t_A / \Delta t_B$ (the ratio of the best-to-worst cycle durations). (**a**) Drug A is more toxic, (**b**) drug B is more toxic. The unrestricted optima for BDF and WDF strategies are marked by black dots; the optima under the toxicity assumptions are marked by stars. The parameters are: $L = 1$, $D = 0$, $H_A = 5$, $H_B = 3$, $u_1 = u_2 = 10^{-5}$, $u_{12} = 100u_1u_2$, $N = 10^{10}$

the objective functions calculated for the two regimes are equal to each other. If for some reason toxicity restrictions require us to use protocols corresponding to the cycle lengths above this line, then a WDF treatment will be preferable.

There are several ways in which the optimization problem can be formulated in the presence of drug side effects. Here, we adopt the following simple approach. We will identify all possible protocols which are acceptable from the point of view of side effects. This is equivalent to separating the $(\Delta t_A, \Delta t_B)$ space into regions of high and low toxicity. Then, we will find the most effective strategy inside the region of low toxicity. Below, we consider two examples of application of this method (Figs. 9.6a, b).

First, let us assume that drug A (the stronger of the drugs) is very toxic. To minimize side effects experienced by the patient, one must minimize (i) the cycle lengths of drug A and (ii) the percentage of time that drug A is administered. Let us denote $\nu = \Delta t_A / \Delta t_B$ (the vertical axes in Fig. 9.6). The above two requirements can be written as follows:

(i) $\Delta t_A < c_1$, where c_1 is some constant. This inequality can be rewritten as $\nu < c_1/\Delta t_B$, and corresponds to the region below the hyperbola in Fig. 9.6a.
(ii) The percentage of time that drug A is administered is given by $\nu/(1 + \nu)$. Requirement $\nu/(1 + \nu) < c_2$, where $c_2 < 1$ is some constant, leads to the inequality $\nu < c_2/(1 - c_2)$. This corresponds to the region below the straight horizontal line in Fig. 9.6a.

From the above considerations, the relatively low toxicity region corresponds to the bottom left corner in Fig. 9.6a. Obviously, the protocol previously identified as optimal cannot be used, and the optimization procedure must happen within the low toxicity region. Since this region is below the thick black line, a BDF protocol will still be the strategy of choice. Minimizing the function F under these constraints, we find the new optimum, which is indicated by a star in Fig. 9.6a. It corresponds to $\Delta t_A = 0.36$ and $\Delta t_B = 0.8$; the percentages of time when drugs A and B are administered are given by 31 and 69 % respectively. As we can see, the new optimum treatment has a shorter cycle duration for drug A and a lower percentage of time when it is used, compared to the old, unrestricted optimum.

In the second example, let us assume that drug B is more toxic than drug A. This imposes the following requirements: (i) the cycle lengths of drug B should be minimized and (ii) the percentage of time that drug B is administered should be restricted. In our notations, this leads to the following restrictions:

(i) $\Delta t_B < c_3$, where c_3 is some constant. This corresponds to the region to the left of a straight vertical line in Fig. 9.6b.
(ii) The percentage of time that drug B is administered is given by $1/(1+v)$. Requirement $1/(1 + v) < c_4$, where $c_4 < 1$ is some constant, leads to the inequality $v > (1 - c_4)/c_4$. This corresponds to the region above the straight horizontal line in Fig. 9.6b.

It follows that the relatively low toxicity region corresponds to the top left corner in Fig. 9.6b. Now this region is above the thick black line, and therefore a WDF strategy is preferable. The new optimum is indicated by a star in Fig. 9.6b, and corresponds to short (mathematically, zero-length) cycles of both drugs, such that the percentage of time when drug B is administered is below 50 %, which is lower than the 58 % of the original optimum.

The examples above describe the following general trend. If the best drug is also more toxic, then the optimal protocol under the toxicity restrictions still starts with the best drug (but uses it for a shorter cycle duration, and at a smaller percentage of time). If however the worst drug is more toxic, then the optimal treatment strategy may start with the worst drug. This is the only case we found where the BDF rule could be overturned. In this case, the better drug must be used for relatively longer durations, and the total cycle durations must be kept short.

To explain this we note that both BDF and WDF optimal strategies require a longer cycle duration for the worst drug; however, the ratio of the best to worst cycle length is smaller in the BDF case. If the best drug is more toxic, we impose a requirement that it should be used at a smaller percentage of time; the corresponding region of the parameter space (the lower left corner) is closer to the BDF optimum, and thus will require a BDF strategy. On the other hand, if the worst drug is more toxic, the low toxicity subspace corresponds to the upper left corner, and may happen in the region where WDF strategies are optimal.

9.6 Summary

In this chapter we studied cyclic drug therapies with the aim to develop general guidelines for optimal treatment scheduling. Our work continues earlier studies on [6] and extends the results to cross-resistant drugs.

We distinguish two different characteristics of the drugs: their *potency* and their *activity spectrum*. Recall that by potency, we mean how effectively a drug kills cells that are susceptible to the drug; this is similar to the killing rate discussed by [6]. Further, a drug is characterized by a broad activity spectrum if it is effective against a large spectrum of mutant cells. On the other hand, a drug with a narrow activity spectrum is a more specific agent, which is active against a relatively small number of cell variants.

One important result is that combination treatments, where both drugs are applied simultaneously, always have a higher probability of treatment success compared to cyclic treatments, and should be used as long as such a combination treatment is tolerated by the patient. This is also supported by clinical data [11]

We find that in order for a cyclic treatment to be effective, the drugs' potencies must satisfy a certain condition (condition (9.15, which we call the condition of "mutual strength"). For realistic parameters, drugs which are not mutually strong will yield very poor probabilities of treatment success, if applied cyclically. The following results are obtained for mutually strong drugs. We find that the "worst drug rule" derived by [6] only holds partially. The general result, which also holds for systems with cross-resistance, states that the optimal strategy is to *use the BDF, but use the worst drug for longer*. For drugs of similar potency and different activity spectra we obtain the rule: *use the less active drug first, and use the more active drug for longer*. We further extend these findings to the case of higher toxicity drugs and show how to optimize treatment in such situations.

Problems

9.1. Research project
In the absence of cross-resistance ($u_{12} = 0$), explore the behavior of the objective function, 9.10, as a function of Δt_B and $\Delta t_A / \Delta t_B$.

9.2. Research project
How does the presence of cross-resistance, see formula (9.11), influence the behavior of the objective function?

9.3. Research project
Can you design a different approach to the optimization of treatment strategies in the context of drugs with high toxicity? How can the information about drug toxicity be included in the optimization function?

9.4. Research project

Find out about the work of Day [6], and about subsequent experimental attempts to verify the "worst drug rule." Were the experiments designed in such a way as to be able to reject/accept the hypothesis of the worst drug rule? In which ways could researchers "miss" it?

9.5. Numerical project

Modify the program obtained for numerical project 4.8 to model the effect of cyclic treatments. Vary the number of cycles and their duration. Experiment with different values of H_1 and H_2 (drug potencies).

References

1. Goldie, J.H., Coldman, A.J.: A mathematic model for relating the drug sensitivity of tumors to their spontaneous mutation rate. Cancer Treat. Rep. **63**(11–12), 1727–1733 (1979)
2. Goldie, J.H., Coldman, A.J., Gudauskas, G.A.: Rationale for the use of alternating non-cross-resistant chemotherapy. Cancer Treat. Rep. **66**, 439–449 (1982)
3. Goldie, J.H., Coldman, A.J.: Quantitative model for multiple levels of drug resistance in clinical tumors. Cancer Treat. Rep. **67**(10), 923–931 (1983)
4. Coldman, A.J., Goldie, J.H.: Role of mathematical modeling in protocol formulation in cancer chemotherapy. Cancer Treat. Rep. **69**(10), 1041–1048 (1985)
5. Goldie, J.H., Coldman, A.J.: Drug Resistance in Cancer: Mechanisms and Models. Cambridge University Press, Cambridge (1998)
6. Day, R.S.: Treatment sequencing, asymmetry, and uncertainty: protocol strategies for combination chemotherapy. Cancer Res. **46**, 3876–3885 (1986)
7. Norton, L., Day, R.: Potential innovations in scheduling of cancer chemotherapy. In: Devita, V.T., Hellman, S., Rosenberg, S.A. (eds.) Important Advances in Oncology, pp. 57–72. Lippincott, Williams & Wilkins, Philadelphia (1985)
8. Day, R.: A branching-process model for heterogeneous cell populations. Math. Biosci. **78**, 73–90 (1986)
9. Katouli, A., Komarova, N.: The worst drug rule revisited: mathematical modeling of cyclic cancer treatments. Bull. math. biol **73**(3), 549–584 (2011)
10. Gaffney, E.A.: The mathematical modelling of adjuvant chemotherapy scheduling: incorporating the effects of protocol rest phases and pharmacokinetics. Bull. Math. Biol. **67**, 563–611 (2005)
11. Shah, N., Skaggs, B., Branford, S., Hughes, T., Nicoll, J., Paquette, R., Sawyers, C., et al.: Sequential abl kinase inhibitor therapy selects for compound drug-resistant bcr-abl mutations with altered oncogenic potency. J. Clin. Invest. **117**(9), 2562 (2007)

Part II
Treatment of cancer with oncolytic viruses

Part II

Chapter 10
Introduction to Oncolytic Viruses

Abstract The second part of the book examines the dynamics of oncolytic viruses, i.e., viruses that have been engineered to specifically infect and kill cancer cells. The aim is for the virus to spread throughout the tumor cell population and to thereby drive the tumor into remission. Healthy cells are not productively infected as viral replication is shut down. Therefore, oncolytic viruses enable us to specifically target cancer cells without having to understand the exact cellular defects that initiate and maintain given tumors. While encouraging treatment results have been reported, consistent treatment success remains elusive. Mathematical models can play an important role, together with experimental studies, in our quest to understand the correlates of treatment success. This part of the book will discuss how mathematical models have been helpful in this respect. The current chapter provides biological background that is important for the modeling.

Keywords Oncolytic viruses · Adenovirus · Virus dynamics · Virus replication · Infected cells · Target cells · Virus transmission · Cancer eradication · Engineered viruses · Ordinary differential equations · Basic reproductive ratio

10.1 What Are Oncolytic Viruses? An Overview

A major challenge with the development of small molecule inhibitors is that it requires a detailed understanding of the molecular defects that drive uncontrolled cellular proliferation. While this has been achieved in the case of CML, our understanding remains relatively poor in the context of other cancers. In many cases, a certain number of mutations need to be accumulated in cells in order to achieve uncontrolled proliferation, and these pathways are not well understood. This complicates the search for effective small molecule inhibitors. The second part of the book will focus on another targeted treatment approach that is less established and at an earlier stage of development: the use of oncolytic viruses [1–17]. Oncolytic viruses

N. L. Komarova and D. Wodarz, *Targeted Cancer Treatment in Silico*,
Modeling and Simulation in Science, Engineering and Technology,
DOI: 10.1007/978-1-4614-8301-4_10, © Springer Science+Business Media New York 2014

Fig. 10.1 Oncolytic viruses
specifically infect cancer
cells, replicate in them, and
transmit the virus to further
tumor cells. The aim is for the
virus to spread through the
tumor cell population and to
eradicate the cancer

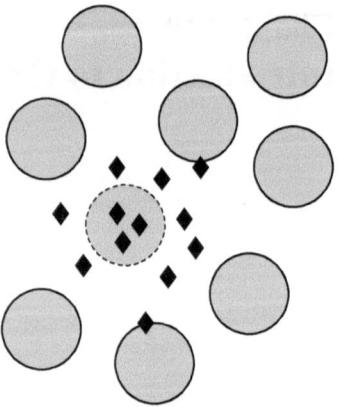

specifically infect cancer cells and replicate in them, spreading from one tumor cell to the other (Fig. 10.1). Importantly, they do not infect healthy cells. While some animal viruses naturally have this characteristic [18], many oncolytic viruses are engineered such that their replication cycle is shutdown by gene products that are present in healthy but absent in cancer cells. The idea is that the virus spreads throughout the tumor cell population, driving it to extinction. This approach is similar in principle to the biological control of agricultural pests [19]. Rather than fighting the pest with chemical agents, natural enemies of the pest can be introduced with the aim to eradicate the pest species. Because of the selectivity of oncolytic viruses for cancer cells rather than normal human cells, side effects are less pronounced than in traditional treatments such as chemotherapy or ionizing radiation. The advantage of this treatment approach is that the detailed molecular defects and pathways that initiate and maintain the cancer do not have to be understood for this treatment to work. The virus can specifically infect cells from a wide variety of cancers as long as certain checkpoints are missing that are commonly deactivated in cancer cells.

Oncolytic virus therapy has been explored in the context of several different virus species, with different viruses being specific for different tissues and tumors [1]. The first engineered virus generated in the 1990s was a herpes simplex virus-1 [20]. Engineered adenoviruses have been of major interest in recent clinical trials, especially in the context of head and neck cancer [15]. Indeed the adenovirus H101(Shanghai Sunway Biotech, Shanghai, China) was approved in China for the treatment of head and neck cancer in combination with chemotherapy [21]. An oncolytic adenovirus is also in clinical trials for glioma in the US [22]. A variety of other virus types have also been explored [23]. However, despite initial promising results and observations in the laboratory and clinic, oncolytic viruses have so far failed to demonstrate sustained and reliable treatment success [14]. Major challenges remain with respect to the method of virus delivery and the ability of the virus to spread through the tumor cell population. It is the latter aspect where mathematical models can be particularly useful for understanding the basic dynamics between a growing tumor cell population and a virus infecting those tumor cells.

10.2 Mathematical Models

Besides experimental research, mathematical and computational modeling has increasingly become a tool to study the dynamics of oncolytic viruses. Mathematical models can help us understand the emerging properties of cancer-virus interactions, to interpret experimental results, and to design new experiments. The first mathematical models of oncolytic virus therapy were developed by Wodarz [24] and considered ordinary differential equations that described the basic interactions between a replicating virus and a growing population of tumor cells, and also immune responses [24, 25]. Further work, both by Wodarz and other groups extended this type of approach in a number of ways, describing different scenarios and applying models to specific virus-tumor systems [26–41], with the aim to refine our understanding of the conditions needed to successfully treat tumors.

Different modeling approaches take into account varying degrees of biological complexity. At one extreme, one could aim to include all biological factors known to influence the spread of the virus and thus the outcome of therapy. It can be argued that such a comprehensive approach is necessary to build predictive models. On the other hand, this approach is limited by many difficulties. Even though we may know the identity of factors that influence virus spread, the exact nature of the interactions, and how to formulate those interactions mathematically, remains unknown. Hence, many aspects of model formulation will be arbitrary, and results/predictions can strongly depend on the exact formulation. A different approach limits the complexity and aims to understand the very basic principles of the dynamics, which then can serve as a foundation to incrementally build in further complexities. Such models often describe simplified situations, such as *in vitro* systems, which can be used to validate the model in a relatively robust fashion. It can be argued that without a basic understanding, it will be impossible to build more complex models that eventually aim to be predictive in an *in vivo* setting. It is in this spirit that we consider mathematical models of virus dynamics, and indeed the whole book follows this principle.

Similar arguments hold for mathematical complexity. Mathematical research on oncolytic viruses currently is at a very early stage, and indeed some of the models described in this book are the first models that were examined to study oncolytic virus dynamics. To do this, relatively simple ordinary differential equation models were adapted from the general field of virus dynamics, where such approaches have been and continue to be very useful, leading to biologically important new insights [42]. Hence, a lot of the chapters describe such approaches, and show how they can be employed to gain important insight into the correlates of successful therapy. We do also consider more complex scenarios, such as spatial and stochastic models. This, however, is also done in the context of relatively simple formulations, such as agent-based models, that can be clearly related to biological data and experimental setups under consideration. In our view, it is very important to build such simple models first and validate them with specific experiments before moving to the analysis of mathematically more complex models, which are more difficult to validate. Such more complex models of oncolytic viruses have been analyzed in the literature, including PDE models [30, 34] and multiscale models [40]. However, due to the uncertainty in assumptions and model formulations, the validity and robustness of results currently remain unclear.

10.3 Virus Dynamics

Our approach falls into the larger field of virus dynamics, which uses mathematical models to study the interactions between viruses and their target cells, as well as immune responses in more complex settings, reviewed e.g., in [42–44]. Basic models of virus dynamics are typically given by ordinary differential equations that describe the average time evolution of viruses and cells. In their simplest form, such models assume a replication cycle that is common to many viruses (Fig. 10.2): infection of cells by free viruses, production of offspring viruses, release of these viruses, followed by further infection of susceptible cells. Denoting the population of susceptible cells by x, the population of infected cells by y, and the free virus population by v, the model is given by the following set of ordinary differential equations.

$$\frac{\mathrm{d}x}{\mathrm{d}t} = \lambda - \mathrm{d}x - \beta'xv, \tag{10.1}$$

$$\frac{\mathrm{d}y}{\mathrm{d}t} = \beta'xv - ay, \tag{10.2}$$

$$\frac{\mathrm{d}v}{\mathrm{d}t} = ky - \delta v. \tag{10.3}$$

The susceptible target cell population is produced with a constant rate λ, dies with a rate d, and becomes infected by virus with a rate β', proportionally to the number

Fig. 10.2 Schematic representation of the basic model of virus dynamics. See text for description

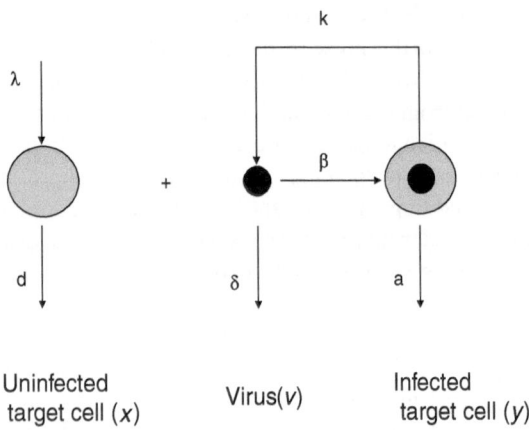

Uninfected target cell (x)   Virus(v)   Infected target cell (y)

Fig. 10.3 Schematic representation of the basic reproductive ratio of the virus, R_0. See text for details

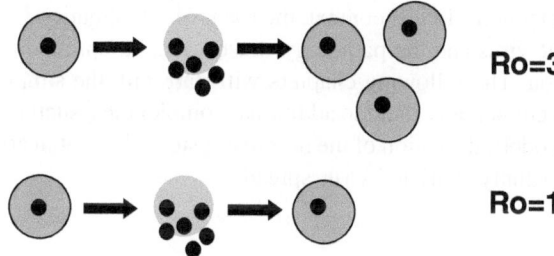

of target cells and the population of free virus. Infected cells produce free virus with a rate k, and free virus decays with a rate δ. The turnover of the virus population is typically much faster than the turnover of infected cells. Therefore, we can assume that the virus population is in a quasi-steady state. Denoting $\beta = \beta'k/\delta$, the model can thus be re-written as

$$\frac{dx}{dt} = \lambda - dx - \beta xy, \tag{10.4}$$

$$\frac{dy}{dt} = \beta xy - ay \tag{10.5}$$

In the most trivial case, the infection is not established and the system converges to the following virus-free equilibrium: $x^{(0)} = \lambda/d$, $y^{(0)} = 0$. In the presence of the infection, the system converges to the following equilibrium describing the coexistence of uninfected and infected cells: $x^{(1)} = a/\beta$, $y^{(1)} = \lambda/a - d/\beta$. The virus can successfully establish an infection if its basic reproductive ratio of greater than one. This is defined as the average number of infected cells produced by a single infected cell when placed into a pool of susceptible cells. It is given by $R_0 = \lambda\beta/da$, and is explained schematically in Fig. 10.3.

10.4 Scope

This model has been much investigated and adapted to describe specific infections of human health importance, such as human immunodeficiency virus (HIV) and hepatitis B and C viruses (HBV and HCV) among others. Focus of the investigation usually is on identifying conditions under which the virus can cause significant pathology and on devising strategies to prevent this. In general, mathematical descriptions of viral dynamics have been vital to understanding the behavior and efficacy of antiviral therapeutics, to our grasp of viral evolution (and evolution in general, for which viruses are among the most important and tractable model systems), and to our understanding of viral pathogenesis and epidemiology [42–46]. The following chapters will explore how this model has been adapted to describe the dynamics of oncolytic virus infections. In this context, the focus of investigation is on enhancing the virulence of the virus and the pathology caused, i.e., on optimizing the virus for tumor destruction. The following chapters will start with the simplest possible models and then incrementally include additional complexities, such as more sophisticated infection models, the action of the immune system, the spatial arrangement of tumor cells, and spatially restricted virus spread.

Problems

10.1. Research project
Find out about the history of oncolytic viruses and the major types of viruses that have been introduced.
10.2. Investigate the behavior of the virus dynamics system (10.4–10.5).
(a) Find the fixed points of system (10.4–10.5).
(b) Perform the linear stability analysis of those equilibria. For what parameter values do we observe stability of each of the solution?
(c) How do the stability properties of the solutions depend on the basic reproductive ratio, $R_0 = \frac{\lambda \beta}{ad}$?

References

1. Russell, S.J., Peng, K.W., Bell, J.C.: Oncolytic virotherapy. Nat. Biotechnol. **30**(7), 658–670 (2012)
2. Breitbach, C.J., De Silva, N.S., Falls, T.J., Aladl, U., Evgin, L., Paterson, J., Sun, Y.Y., Roy, D.G., Rintoul, J.L., Daneshmand, M., Parato, K., Stanford, M.M., Lichty, B.D., Fenster, A., Kirn, D., Atkins, H., Bell, J.C.: Targeting tumor vasculature with an oncolytic virus. Mol. Ther. **19**(5), 886–894 (2011)
3. Bell, J.C.: Oncolytic viruses: what's next? Curr. Cancer Drug Targets **7**(2), 127–131 (2007)
4. Bell, J.C., Lichty, B., Stojdl, D.: Getting oncolytic virus therapies off the ground. Cancer Cell **4**(1), 7–11 (2003)
5. Crompton, A.M., Kirn, D.H.: From onyx-015 to armed vaccinia viruses: the education and evolution of oncolytic virus development. Curr. Cancer Drug Targets **7**(2), 133–139 (2007)
6. Davis, J.J., Fang, B.: Oncolytic virotherapy for cancer treatment: challenges and solutions. J. Gene Med. **7**(11), 1380–1389 (2005)
7. Kaplan, J.M.: Adenovirus-based cancer gene therapy. Curr. Gene Ther. **5**(6), 595–605 (2005)
8. Kelly, E., Russell, S.J.: History of oncolytic viruses: genesis to genetic engineering. Mol. Ther. **15**(4), 651–659 (2007)
9. Kirn, D., Hermiston, T., McCormick, F.: ONYX-015: clinical data are encouraging. Nat. Med. **4**(12), 1341–1342 (1998)
10. O'Shea, C.C.: Viruses—seeking and destroying the tumor program. Oncogene **24**(52), 7640–7655 (2005)
11. Parato, K.A., Senger, D., Forsyth, P.A., Bell, J.C.: Recent progress in the battle between oncolytic viruses and tumours. Nat. Rev. Cancer **5**(12), 965–976 (2005)
12. Post, D.E., Shim, H., Toussaint-Smith, E., Van Meir, E.G.: Cancer scene investigation: how a cold virus became a tumor killer. Future Oncol. **1**(2), 247–258 (2005)

13. Roberts, M.S., Lorence, R.M., Groene, W.S., Bamat, M.K.: Naturally oncolytic viruses. Curr. Opin. Mol. Ther. **8**(4), 314–321 (2006)
14. Wong, H.H., Lemoine, N.R., Wang, Y.: Oncolytic viruses for cancer therapy: overcoming the obstacles. Viruses **2**, 78–106 (2010)
15. Vaha-Koskela, M.J., Heikkila, J.E., Hinkkanen, A.E.: Oncolytic viruses in cancer therapy. Cancer Lett **254**, 178–216 (2007)
16. McCormick, F.: Cancer-specific viruses and the development of ONYX-015. Cancer Biol. Ther. **2**(4 Suppl 1), S157–S160 (2003)
17. McCormick, F.: Future prospects for oncolytic therapy. Oncogene **24**(52), 7817–7819 (2005)
18. Kirn, D.H., McCormick, F.: Replicating viruses as selective cancer therapeutics. Mol. Med. Today **2**(12), 519–527 (1996)
19. Waage, J.K., Hassell, M.P.: Parasitoids as biological-control agents—a fundamental approach. Parasitology **84**(Apr), 241–268 (1982)
20. Martuza, R.L., Malick, A., Markert, J.M., Ruffner, K.L., Coen, D.M.: Experimental therapy of human glioma by means of a genetically engineered virus mutant. Science **252**(5007), 854–856 (1991)
21. Garber, K.: China approves world's first oncolytic virus therapy for cancer treatment. J. Natl. Cancer Inst. **98**(5), 298–300 (2006)
22. Jiang, H., Gomez-Manzano, C., Lang, F.F., Alemany, R., Fueyo, J.: Oncolytic adenovirus: preclinical and clinical studies in patients with human malignant gliomas. Curr. Gene Ther. **9**(5), 422–427 (2009)
23. Eager, R.M., Nemunaitis, J.: Clinical development directions in oncolytic viral therapy. Cancer Gene Ther. **18**(5), 305–317 (2011)
24. Wodarz, D.: Viruses as antitumor weapons: defining conditions for tumor remission. Cancer Res. **61**(8), 3501–3507 (2001)
25. Wodarz, D.: Gene therapy for killing p53-negative cancer cells: use of replicating versus non-replicating agents. Hum. Gene Ther. **14**(2), 153–159 (2003)
26. Bajzer, Z., Carr, T., Josic, K., Russell, S.J., Dingli, D.: Modeling of cancer virotherapy with recombinant measles viruses. J. Theor. Biol. **252**(1), 109–122 (2008)
27. Biesecker, M., Kimn, J.H., Lu, H., Dingli, D., Bajzer, Z.: Optimization of virotherapy for cancer. Bull. Math. Biol. **72**(2), 469–489 (2010)
28. Dingli, D., Cascino, M.D., Josic, K., Russell, S.J., Bajzer, Z.: Mathematical modeling of cancer radiovirotherapy. Math. Biosci. **199**(1), 55–78 (2006)
29. Dingli, D., Offord, C., Myers, R., Peng, K.W., Carr, T.W., Josic, K., Russell, S.J., Bajzer, Z.: Dynamics of multiple myeloma tumor therapy with a recombinant measles virus. Cancer Gene Ther. **16**(12), 873–882 (2009)
30. Friedman, A., Tian, J.P., Fulci, G., Chiocca, E.A., Wang, J.: Glioma virotherapy: effects of innate immune suppression and increased viral replication capacity. Cancer Res. **66**(4), 2314–2319 (2006)
31. Karev, G.P., Novozhilov, A.S., Koonin, E.V.: Mathematical modeling of tumor therapy with oncolytic viruses: effects of parametric heterogeneity on cell dynamics. Biol. Direct **1**, 30 (2006)
32. Komarova, N.L., Wodarz, D.: Ode models for oncolytic virus dynamics. J. Theor. Biol. **263**(4), 530–543 (2010)
33. Novozhilov, A.S., Berezovskaya, F.S., Koonin, E.V., Karev, G.P.: Mathematical modeling of tumor therapy with oncolytic viruses: regimes with complete tumor elimination within the framework of deterministic models. Biol. Direct **1**, 6 (2006)
34. Wein, L.M., Wu, J.T., Kirn, D.H.: Validation and analysis of a mathematical model of a replication-competent oncolytic virus for cancer treatment: implications for virus design and delivery. Cancer Res. **63**(6), 1317–1324 (2003)
35. Wodarz, D.: Computational approaches to study oncolytic virus therapy: insights and challenges. Gene Ther. Mol. Biol. **8**, 137–146 (2004)
36. Wodarz, D.: Use of oncolytic viruses for the eradication of drug-resistant cancer cells. J. R. Soc. Interface **6**(31), 179–186 (2009)

37. Wodarz, D., Komarova, N.: Towards predictive computational models of oncolytic virus therapy: basis for experimental validation and model selection. PLoS ONE **4**(1), e4271 (2009)
38. Zurakowski, R., Messina, M.J., Tuna, S.E., Teel, R.A.: HIV treatment scheduling via robust nonlinear model predictive control. 5th Asian Control Conference 2004 (2004)
39. Mok, W., Stylianopoulos, T., Boucher, Y., Jain, R.K.: Mathematical modeling of herpes simplex virus distribution in solid tumors: implications for cancer gene therapy. Clin. Cancer Res. **15**(7), 2352–2360 (2009)
40. Paiva, L.R., Binny, C., Ferreira Jr., S.C., Martins, M.L.: A multiscale mathematical model for oncolytic virotherapy. Cancer Res. **69**(3), 1205–1211 (2009)
41. Reis, C.L., Pacheco, J.M., Ennis, M.K., Dingli, D.: In silico evolutionary dynamics of tumour virotherapy. Integr. Biol. (Camb.) **2**(1), 41–45 (2010)
42. Nowak, M.A., May, R.M.: Virus Dynamics. Mathematical Principles of Immunology and Virology. Oxford University Press, Oxford (2000)
43. Perelson, A.S.: Modelling viral and immune system dynamics. Nat. Rev. Immunol. **2**(1), 28–36 (2002)
44. Wodarz, D.: Killer Cell Dynamics: Mathematical and Computational Approaches to Immunology. Springer, New York (2006)
45. Anderson, R.M., May, R.M.: Infectious Diseases of Humans. Oxford University Press, Oxford (1991)
46. De Boer, R.J., Oprea, M., Antia, R., Murali-Krishna, K., Ahmed, R., Perelson, A.S.: Recruitment times, proliferation, and apoptosis rates during the CD8(+) T-cell response to lymphocytic choriomeningitis virus. J. Virol. **75**(22), 10663–10669 (2001)

Chapter 11
Basic Dynamics of Oncolytic Viruses

Abstract This chapter adopts general virus dynamics models to describe the basic interactions between a growing tumor cell population and an oncolytic virus that replicates in these tumor cells. Model properties are explored, and the correlates of treatment success that are predicted by the model are discussed. For the virus to control or eradicate the tumor cell population, it needs to spread sufficiently fast. One aspect that determines the viral spread rate in the model is the virus-induced death rate of infected cells. If the virus kills cells too quickly, then the total amount of virus produced by the infected cell becomes lower, and this can prevent the virus from controlling the tumor. Hence, viruses should not be engineered to maximize cell killing, as has been previously suggested. We compare the administration of replicating versus nonreplicating viruses, and discuss what types of experimental assays should be performed to evaluate potential candidate viruses in culture.

Keywords Oncolytic viruses · Virus dynamics · Ordinary differential equations · Infected and target cells · Virus-induced death rate · Tumor control · Virus cytopathicity · MOI · Adenovirus vectors · Virus replication kinetics · Animal models · Engineered viruses

11.1 Introduction

Chapter 10 has given an overview and introduction to the biology of oncolytic viruses, highlighting the complex nature of the interactions between the viruses, their target cells, and the in vivo environment. Certain virus treatment approaches have shown promising results, while other case studies resulted in failure. Even if initial success was observed, control tended to be lost in the long term. Experimental work and clinical trials are based on verbal and graphical reasoning and explore treatment regimes based on verbal hypotheses. The complexities of the underlying interactions, however, do not allow verbal and graphical reasoning to provide a full understanding

N. L. Komarova and D. Wodarz, *Targeted Cancer Treatment in Silico,*   147
Modeling and Simulation in Science, Engineering and Technology,
DOI: 10.1007/978-1-4614-8301-4_11, © Springer Science+Business Media New York 2014

or accurate prediction of the outcome of these interactions, thus limiting the power of this approach. Mathematical and computational models comprise an essential tool that formulates a set of biologically important assumptions and follows them to their precise logical conclusions. They allow us to obtain insights into the predicted outcomes of complex interactions, which can often be counter intuitive and not obtainable by other techniques. The insights obtained from the mathematical models allow us to interpret experimental data, estimate parameters, identify what needs to be measured, and to provide logical guidelines according to which the therapeutic approaches can be further improved and developed. This chapter describes the simplest and most basic of such models and highlights important biological insights that have been obtained. Subsequent chapters will build on this and introduce further biological complexity.

11.2 The Model

This section introduces a simple mathematical model describing the development of a growing tumor and an oncolytic virus population over time. The model is based on extensive general literature on virus dynamics, e.g., see [1–3], and includes three variables: the growing cancer cells, x, infected cancer cells, y, and virus particles, v. It is described schematically in Fig. 11.1 and is given by the following set of ordinary differential equations which describe the development of these populations over time:

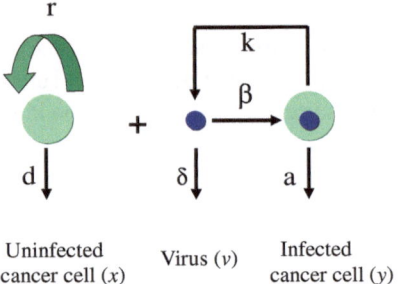

Fig. 11.1 Schematic illustration of the mathematical model which describes the basic dynamics of tumor-cell infecting viruses. Uninfected cells "react" with free virus to give rise to infected cells; the rate constant is β. Infected cells produce free virions at a rate k. Uninfected cells, free virus, and infected cells die at the rates d, δ, and a, respectively. Uninfected cells are replenished by proliferation at a rate r. For further details, see text

$$\frac{dx}{dt} = rx\left(1 - \frac{x+y}{\omega}\right) - dx - \beta'xv, \tag{11.1}$$

$$\frac{dy}{dt} = \beta xv - (d+\alpha)y, \tag{11.2}$$

$$\frac{dv}{dt} = ky - \delta v. \tag{11.3}$$

The cancer cells grow at a rate r, and this growth is density dependent, limited by a maximum size ω. In biological terms, this means that the cancer cells divide and that this results in exponential growth at small tumor cell densities, but that growth is slowed down as the tumor reaches larger sizes and runs out of space, nutrients, and other resources required for growth. Cancer cells die at a rate d. Therefore the average life span of the cancer cells is given by $1/d$. The cancer cells become infected by the virus at a rate β'. The rate constant, β', describes the efficacy of this process, including the rate at which virus particles find uninfected cells, the rate of virus entry, and the rate of successful infection. Infected cancer cells also have a death rate. The death rate of infected cells, a, is a composite of the natural death rate, d, and the virus-inflicted death rate, α: $a = d + \alpha$. Therefore the average lifetime of an infected cell is $1/(d + \alpha)$. Infected cells produce virus at a rate k. The total amount of virus particles produced from one infected cell, or the "burst size", is hence given by $k/(\alpha+d)$. Finally, the virus decays at a rate δ. Thus, the average life span of a virus particle is given by $1/\delta$. In the context of replicating versus nonreplicating viruses, we can make the following distinction. A nonreplicating virus is characterized by $k = 0$ (no virus production by the infected cell), while a replicating virus is characterized by $k > 0$.

In this model, the tumor expands if its growth rate is greater than its death rate, i.e. if $r > d$. In the absence of treatment, the tumor will eventually grow to its maximum size given by $x^{(0)} = \omega(r - d)/r$. Treatment in the model corresponds to the introduction of virus into the system. The virus has the potential to spread if $\beta x^{(0)} > d + \alpha$, that is, if the replication rate of the virus is fast relative to the death rate of the infected cancer cells. The virus infection then takes the system to a new equilibrium outcome which is given by the following expressions:

$$x^{(1)} = (d + a)/\beta, \tag{11.4}$$

$$y^{(1)} = [\beta\omega(r - d) - r(d + \alpha)]/[\beta(r + \beta\omega)], \tag{11.5}$$

$$v^{(1)} = ky^{(1)}/\delta, \tag{11.6}$$

where β summarizes the overall replication rate of the virus and is given by $\beta = \beta'k/\delta$. The size of the overall tumor cell population during virus therapy is given by the sum of uninfected and infected tumor cells, $x^{(1)} + y^{(1)}$. The aim of therapy should be to reduce this population to low levels. If it has been reduced below a threshold in the model, the tumor population can be assumed to be extinct (number of tumor cells below one).

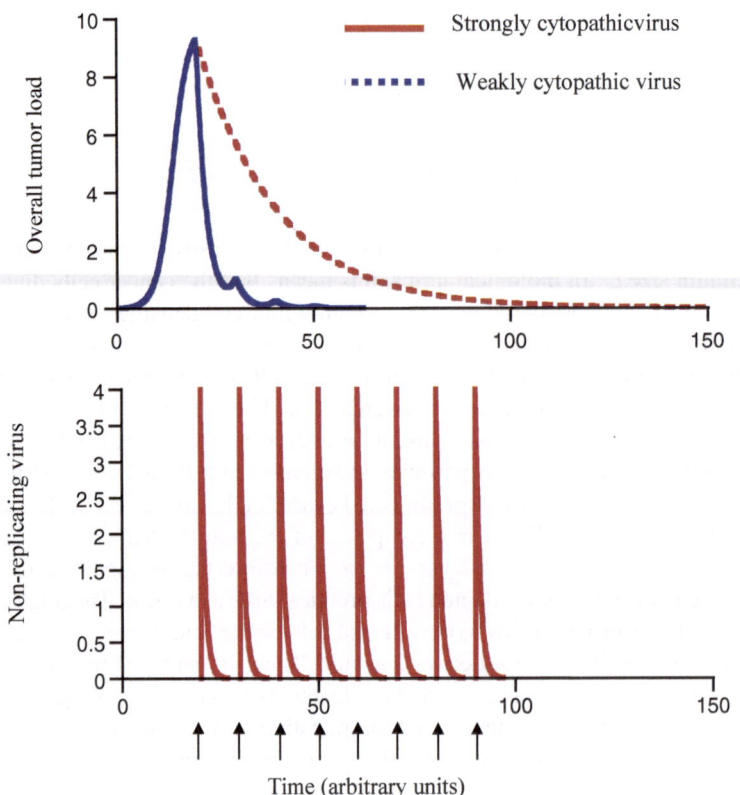

Fig. 11.2 Simulation of therapy using a nonreplicating virus. The virus is administered repeatedly, as indicated by the arrows. The development of tumor load over time is shown assuming both a strongly cytopathic virus and a weakly cytopathic virus. The strongly cytopathic virus results in more efficient eradication of the tumor. Parameters were chosen as follows. $r = 0.5$, $\omega = 10$, $\beta' = 1.5$; $d = 0.01$; $k = 0$; $\delta = 1$. For the strongly cytopahtic virus, $\alpha = 0.4$. For the weakly cytopathic virus, $\alpha = 0.04$

11.3 Nonreplicating Viruses

Assume that the virus is not replicating ($k = 0$). If not all cancer cells become infected after the first administration, it has to be given repeatedly in order to ensure continued presence of the virus and hence remission of the cancer (Fig. 11.2). The goal is to eradicate the population of uninfected tumor cells. Since infected cancer cells do not divide, they do not pose a threat and will decay to extinction. The dynamics of the uninfected cancer cell population over time during treatment can be approximated by

$$x(t) = x^{(0)}e^{(r-d-\beta'v)t}.$$

This assumes that the tumor is still growing and has not yet reached levels close to maximum tumor size ($x \ll \omega$). For the uninfected cancer cells to decline, the level of virus has to be kept above a threshold during therapy, given by $v > (r - d)/\beta$. Thus, a high infectivity of the virus, and a low growth rate of the tumor facilitate tumor reduction. If this condition is fulfilled, the half life of the uninfected tumor cell population is given by

$$t_{1/2} = \ln(1/2)/(r - d - \beta'v).$$

The time to eradication of the uninfected tumor cells is hence given by

$$t = \frac{\ln(x_0)}{\ln(2)} \frac{\ln(1/2)}{r - d - \beta'v}.$$

After this time threshold, administration of the virus can be stopped, and the population of infected tumor cells decays with a half life of $t_{1/2} \ln 2/a$. Hence, the faster the rate of virus-induced cell death, α, the faster the population of infected cells declines toward extinction (Fig. 11.2). To summarize, the best strategy is to (1) use a cytopathic virus, (2) use a virus with high infectivity, and (3) reduce the cancer growth rate which might be achieved by certain chemo- or radio-therapeutic regimes [4].

11.4 Replicating Viruses

Now assume a replicating virus $k > 0$ If the patient is injected with a very high inoculum dose of the virus, most or all cancer cells immediately become infected and the situation is the same as for the nonreplicating virus. If, however, the initial virus inoculum is not that large and does not immediately overwhelm the cancer, then the dynamics between virus replication and tumor growth will play out, resulting in an equilibrium outcome. The size of the tumor load (infected + uninfected cells) at equilibrium determines the level of success. If equilibrium tumor load in the model is very low, this corresponds to eradication in practical terms. Higher equilibrium tumor loads in the model correspond to persistence of the tumor in the presence of the virus. Equilibrium tumor load is given by

$$x^{(1)} + y^{(1)} = \frac{\omega(\alpha + r)}{r + \beta\omega}.$$

An important result is that a lower virus-induced death rate of infected cells (small a) results in lower equilibrium tumor load (Fig. 11.3). If the rate of virus-induced tumor cell killing is too high, the outcome is persistence of the tumor in the face of ongoing viral replication (Fig. 11.3). The reason is as follows. Low viral cytopathicity increases virus load. Higher virus load results in more infection and in a greater

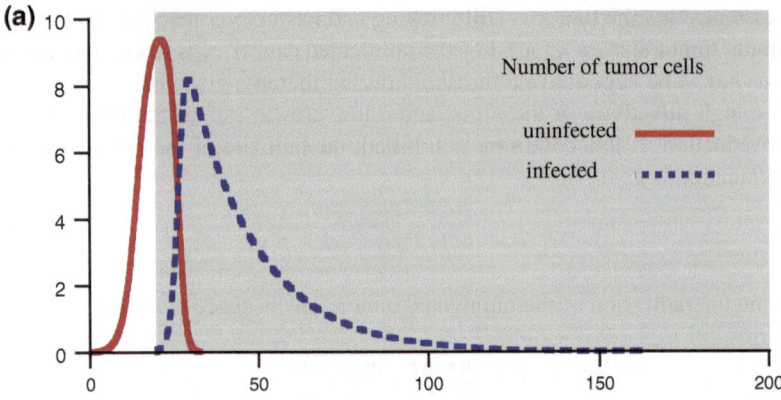

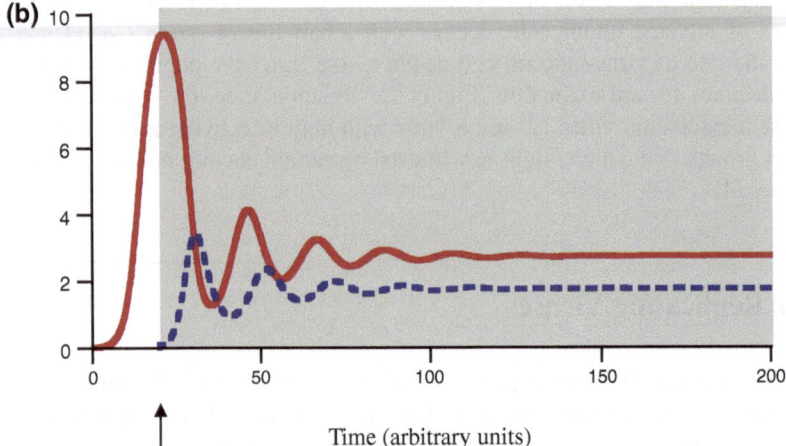

Time (arbitrary units)

Fig. 11.3 Simulation of therapy using a replicating virus. The virus is administered once, as indicated by the arrow. Shading indicates the phase of the dynamics following administration of the virus. (**a**) Use of a weakly cytopathic virus results in sustained cancer remission. (**b**) Use of a more cytopathic virus results in long-term persistence of the cancer and the virus. Parameters were chosen as follows: $r = 0.5$; $\omega = 10$; $\beta' = 1.5$; $d = 0.01$; $k = 0.1$; $\delta = 1$. For (**a**) $\alpha = 0.04$. For (**b**) $\alpha = 0.4$

decline of the uninfected cancer cells. Higher viral cytopathicity results in lower virus load. Low virus load results in less infection and in less reduction of the uninfected tumor cells. In addition, equilibrium tumor load is reduced by other parameters, most notably by a high replication rate of the virus (high value of β), and a slow growth rate of the tumor (low value of r). As expected, the replication rate of the virus (value of β) has to lie above a threshold for tumor eradication to be possible. If the replication rate of the virus is too low, the virus does not spread sufficiently through the population of tumor cells. In summary, if the virus is replicating, the best strategy is to (1) use a weakly cytopathic virus, (2) use a fast replicating virus, and (3) reduce the growth rate of the tumor by alternative therapeutic means.

11.5 Evaluation of Replicating Viruses in Culture

The mathematical model has given rise to an important difference in treatment strategy depending on whether the virus replicates or not. If the virus does not replicate, a high degree of cytopathicity is beneficial because it speeds up elimination of all tumor cells. On the other hand, if the virus replicates, success is promoted by using a weakly cytopathic virus. A high rate of virus-induced cell death is detrimental and leads to the persistence of both tumor and virus. These findings also have important implications for the methods used to evaluate potential viruses in culture. If the virus does not replicate, a high multiplicity of infection (MOI) has to be used. The virus with the strongest degree of tumor cell killing will remove the cancer cells fastest. Such a virus will also work best in the physiological situation, since the aim is to overwhelm the tumor by repeatedly injecting the agent, resulting in fast killing of as many cancer cells as possible (Fig. 11.2). On the other hand, if the virus replicates, a low MOI is required to evaluate the virus. The reason is that *in vivo*, the replicating virus has to spread through the cancer cell population, and this has to be mimicked in culture. Using a high MOI can result in misleading evaluations. These notions are illustrated in Fig. 11.4 with computer simulations. This figure depicts the dynamics in culture for strongly and weaky cytopathic viruses, using different MOIs. Figure 11.4(a) shows the dynamics for a high MOI. In this simulation, the strongly cytopathic virus results in quick elimination of the tumor cells, while the weakly cytopathic virus is much less effective. Thus, if viruses are evaluated using a high MOI, the virus with the strongest degree of tumor cell killing receives the highest grades. Importantly, this is the virus which is predicted to be least efficient at reducing tumor load *in vivo*. The situation is different when viruses are evaluated in culture using a low MOI (Fig. 11.4b). The less cytopathic virus results in elimination of tumor cells in culture, while the more cytopathic virus fails to eliminate tumor cells in culture. Therefore the less cytopathic virus gets the better marks, and this is also the virus which is predicted to be more efficient at reducing tumor load *in vivo*.

11.6 Implications for Interpreting Data

It is interesting to consider experimental data in the light of the theoretical results. Harrison et al. [5] have used A549 human lung cancer nude mouse xenografts as a model system to study the efficacy of replicating adenovirus vectors at reducing tumor growth. They considered both a wild-type adenovirus (Ad309) and an E1b-19kD-deleted virus (Ad337). While the Ad337 virus was generally more efficient at reducing tumor growth than the wild-type Ad309, both viruses failed to eradicate

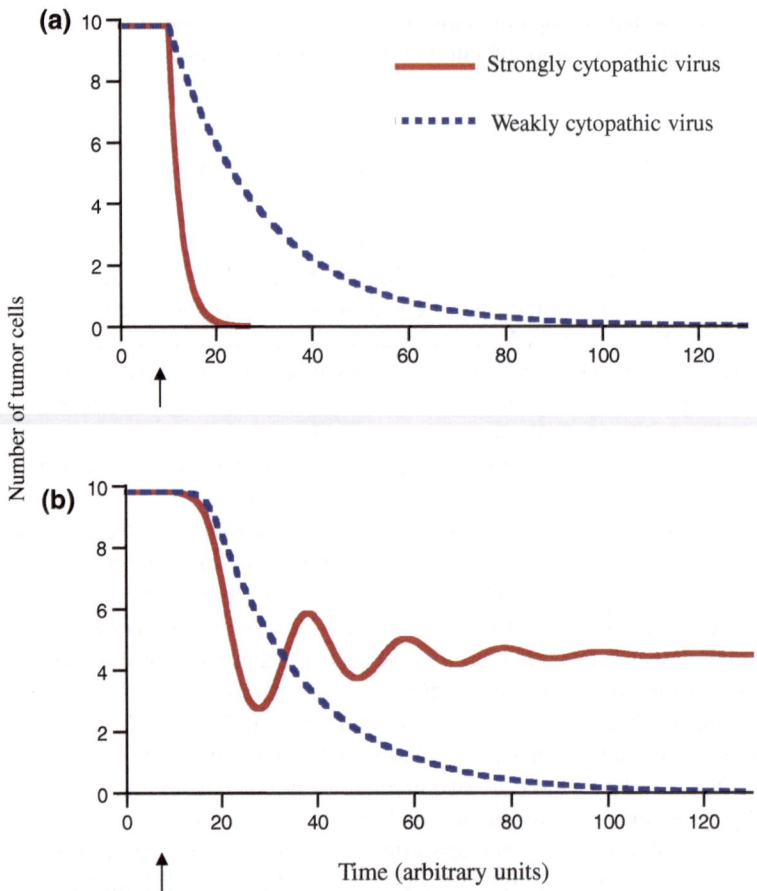

Fig. 11.4 Simulation showing the evaluation of potential replicating viruses in culture. A weakly and a strongly cytopathic virus are compared when the culture is inoculated (see arrow) with (**a**) a high MOI and (**b**) a low MOI. While under high MOI infection the more cytopathic virus yields better results, the opposite is true for low MOI infection, which reflects virus spread conditions in vivo. Parameters were chosen as follows. $r = 0.5; \omega = 10; \beta' = 1.5; d = 0.01; k = 0.1; \delta = 1$. For the strongly cytopahtic virus, $\alpha = 0.4$. For the weakly cytopathic virus, $\alpha = 0.04$. Virus inoculum was $v = 10$ for high MOI and $v = 0.01$ for low MOI

established tumors. Instead, the tumor persisted in the presence of ongoing viral replication. The reason for the failure to eradicate the tumor despite ongoing viral replication was left open to speculation. Several mechanisms have been suggested [5], most prominently the presence of physical barriers within the tumor which can limit viral spread (such as connective tissue and tumor matrix). This argument was further supported by the observation that when tumor cells were infected in vitro and mixed with uninfected tumor cells, 1 % or fewer infected cells were required to prevent tumor establishment. The modeling results presented here add to this discussion. It might not so much be the physical barriers that contribute to the failure of the virus to eliminate the tumor. In the light of theory, it is also possible that the virus is too cytopathic. That is, the rate of virus-mediated destruction of the cancer cells is too high. This limits the overall replicative potential of the virus population, resulting in persistence of both the virus and the tumor. If the virus is mixed early with the tumor, before it has become established, it is easier for the virus to infect and kill all the cells before the cancer gets a chance to grow. These arguments are also supported by experiments using an attenuated herpes simplex virus-1 (HSV-1) [6]. A $\gamma 34.5$ mutant virus was modified by genetic selection to spread more efficiently through tumors. While the selected virus was more efficient than the parent virus, this was most obvious at reduced viral doses relative to the size of the tumor. Selection is likely to have resulted in an increased replication rate of the virus, or a decreased rate of cell killing. Both traits increase the overall spread of the virus through the tumor, and promote tumor eradication.

11.7 Summary

To summarize, in contrast to nonreplicating agents, it might not be a good strategy to maximize the "oncolytic" potential of replicating viruses. Instead, the ability of the virus to spread through the tumor should be maximized, and this is promoted by a slow rate of cell killing. In order to design better replicating viral vectors, mathematical models should be coupled with experimental data to measure the relevant parameters. In particular, it will be important to know what it means that the virus can bee "too lytic" and thus prevent success. In practical terms, there will be a threshold level of cytopathicity: above the threshold, overall spread of the virus is compromised to the degree that we observe failure of therapy. Below this threshold, therapy can result in sustained tumor remission. The exact level of this cytopathicity threshold depends on many factors, most importantly the rate of viral replication and the size of the tumor. It will be important to precisely measure the rate at which different viruses kill the tumor cells. This can be done in the following way. Infect a culture of cells with a high MOI. This will quickly result in the infection of most or all cells in culture. Because in this situation, there will not be any further tumor cells to infect, the virus dynamics will be governed by the exponential decay of infected cells. In the model, this corresponds to $dy/dt = -(\alpha + d)y$. The slope of the exponential decay on a log axis gives an estimate of the death rate of virus infected tumor cells. This rate of virus-induced killing of infected cells should then be correlated with the efficacy at which the virus reduces the size of the tumor cell population, starting with a low MOI. This should be done both *in vitro* and *in vivo* (using animal models in which the initial tumor size can be varied).

Besides the rate of virus-induced cell killing, it will also be important to measure the replication kinetics of the virus. As pointed out, if the replication kinetics (value of β') are not sufficiently high, virus therapy cannot eradicate the cancer. The viral replication kinetics can be measured as follows *in vitro*. Infect a dish containing a certain number of tumor cells with a low MOI and monitor the growth of the virus population closely. Initially, exponential growth of the virus population is observed which subsequently slows down as the virus runs out of cells to infect. The exponential phase of this growth is described by the differential equation $dy/dt = \beta x^* y - (\alpha + d)y$, where x^* is the number of tumor cells in culture. The slope of this exponential increase of the virus population on a log axis gives the intrinsic reproductive rate of the virus $(\beta x^* - (\alpha + d))$. If the initial number of tumor cells, and the rate of virus-induced cell killing are known, we can obtain a measure of the replication rate parameter β.

Problems

11.1 Numerical project
Further examine the properties of the model given by Eqs. (11.1–11.3), using both analytical and numerical methods. In particular, examine how the outcome of virus treatment depends on the growth rate of the tumor cell population, r. This growth rate can potentially be modified by drug therapies. Would it be beneficial to combine virus treatment with drug therapy?

11.2 Numerical project
Examine analytically and numerically, how the properties of model (11.1–11.3) change if you assume that the death rate of infected cells correlates with the replication rate of the virus through the following dependency: $a = f\beta$, where $f > 0$ is a constant factor.

11.3 Research project
Some oncolytic viruses are engineered to carry death or suicide genes. Find out about these approaches. According to the models discussed here, do such approaches enhance or limit the success of treatment?

References

1. Nowak, M.A., May, R.M.: Virus Dynamics: Mathematical Principles of Immunology and Virology. Oxford University Press, New York (2000)
2. Wodarz, D.: Killer Cell Dynamics: Mathematical and Computational Approaches to Immunology. Springer, New York (2006)
3. Perelson, A.S.: Modelling viral and immune system dynamics. Nature Rev. Immunol. 2(1), 28–36 (2002)
4. You, L., He, B., Xu, Z., McCormick, F., Jablons, D.M.: Future directions: oncolytic viruses. Clin. Lung Cancer 5(4), 226–230 (2004)
5. Harrison, D., Sauthoff, H., Heitner, S., Jagirdar, J., Rom, W.N., Hay, J.G.: Wild-type adenovirus decreases tumor xenograft growth, but despite viral persistence complete tumor responses are rarely achieved- deletion of the viral e1b–19-kd gene increases the viral oncolytic effect. Hum. Gene. Ther. 12(10), 1323–1332 (2001)
6. Taneja, S., MacGregor, J., Markus, S., Ha, S., Mohr, I.: Enhanced antitumor efficacy of a herpes simplex virus mutant isolated by genetic selection in cancer cells. Proc. Nat.l Acad. Sci. USA 98(15), 8804–8808 (2001)

Chapter 12
Mitotic Virus Transmission and Immune Responses

Abstract This chapter builds on the mathematical models of oncolytic virus dynamics discussed in the previous chapter and introduces further biological complexity. In extending of the model, we allow for mitotic transmission of the virus, i.e. virus spread through cell division. Depending on the rate at which infected cells divide, we can observe an optimal rate of virus-induced cell killing that minimizes tumor load. Rates of cell killing that are smaller or larger than the optimum lead to higher tumor loads during treatment. In another extension of the model, we introduce immune responses. Anti-viral immune responses can kill infected cells (and thus influence their death rate). Under this assumption, we find an optimal rate of immune-induced cell killing that minimizes tumor load. Less effective immune responses can lead to higher tumor loads through less killing of cells, and too strong of an immune response significantly impairs or eliminates the virus. In addition, a tumor specific immune response is considered that can be activated through virus-induced necrosis of cells. In this case, the effect of virus therapy is augmented and an increase in the strength of anti-tumor immunity is always beneficial for therapy.

Keywords Anti-viral immune responses · Oncolytic viruses · Viral transmission pathways · Mitotic transmission · Vertical transmission · Ordinary differential equations · Tumor antigen · Optimal viral cytotoxicity · Tumor load · Virus-specific CTL · Tumor-specific CTL

12.1 Introduction

Beyond the basic dynamics described in the previous chapter, a multitude of further biological factors can come into play and influence the dynamics. This chapter concentrates on two additional aspects and demonstrates the rather complex dynamics that result. First, we will consider an additional viral transmission pathway. Besides transmission of the virus from cell to cell, the virus can also potentially spread via the division of infected cells. This may be unrealistic for viruses that arrest the cells in the S-phase of the cell cycle, but can apply to other viruses. This can be called "mitotic transmission" or "vertical transmission" [1]. Second, we will consider a factor that certainly plays a significant role in vivo, i.e. immune responses. In this context, we distinguish between two types of immune responses: anti-viral immune responses that recognize and kill virus-infected tumor cells, and tumor-specific immune responses that recognize tumor antigens on the surface of tumor cells (uninfected or infected) and kill the cells. It is thought that tumor-specific immune responses are often silent because tumor cells are "self" and behave like this in many ways, importantly not causing a lot of necrotic cell death that can trigger signals that activate anti-tumor immunity [2, 3]. Most cell death in tumors is likely apoptotic, which tends not to activate those signals. A virus infection can lead to significant amounts of necrotic cell death, and could thus trigger the activation of anti-tumor immune responses, which in turn could have a significant impact on the outcome of treatment [4–6]. These assumptions will be built into the basic model considered in the previous chapters, and it will be investigated how model properties are influenced.

12.2 Effect of Mitotic or Vertical Transmission of the Virus

In this section we elaborate on the model introduced in the previous chapter and allow for the division of infected cells. Whether this applies or not depends on the specific virus in question, and the extent of infected cell proliferation is determined by a model parameter and can be varied from low to high. The expanded model contains two variables: uninfected tumor cells, x, and tumor cells infected by the virus, y. The free virus population is assumed to be in quasi-steady state and is not explicitly modeled in this formulation. This is possible because the turnover of the virus population is much faster than that of the infected cells. The model is explained schematically in Fig. 12.1.

The tumor cells grow in a logistic fashion at a rate r and die at a rate d. The maximum size or space the tumor is allowed to occupy is given by its carrying capacity ω. The virus spreads to tumor cells at a rate β (this parameter can be viewed as summarizing the replication rate of the virus). Infected tumor cells are killed by the virus at a rate a and grow in a logistic fashion at a rate s. This assumes that division of infected tumor cells results in both daughter cells carrying the virus. This would certainly be the case with a virus that integrates into the tumor cell genome, but even with a non-integrating virus, the chances of transmission upon cell division should be sufficiently high to justify this assumption. The model is given by the following set of ordinary differential equations [7].

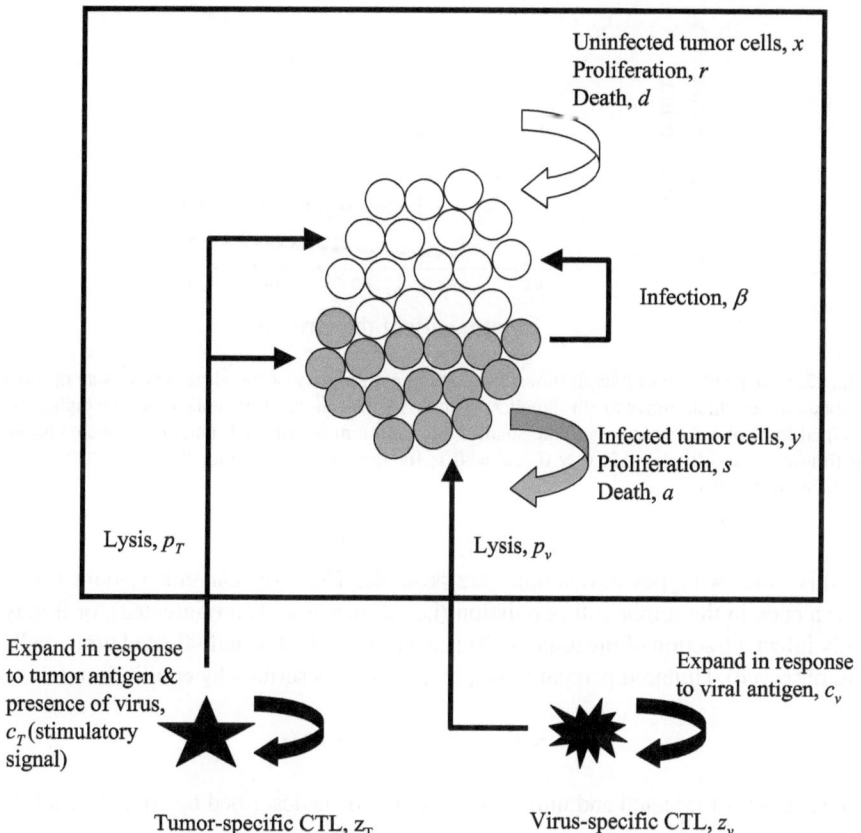

Fig. 12.1 Schematic representation of the models which are reviewed in this chapter

$$\dot{x} = rx\left(1 - \frac{x+y}{\omega}\right) - dx - \beta xy, \qquad (12.1)$$

$$\dot{y} = \beta xy + sy\left(1 - \frac{x+y}{\omega}\right) - ay. \qquad (12.2)$$

In the absence of the virus the trivial equilibrium is attained and is given by E0:

$$x^{(0)} = \omega(r - d)/r, \quad y^{(0)} = 0.$$

The virus can establish an infection in the tumor cell population if

$$\frac{\beta\omega(r - d) + sd}{r} > a.$$

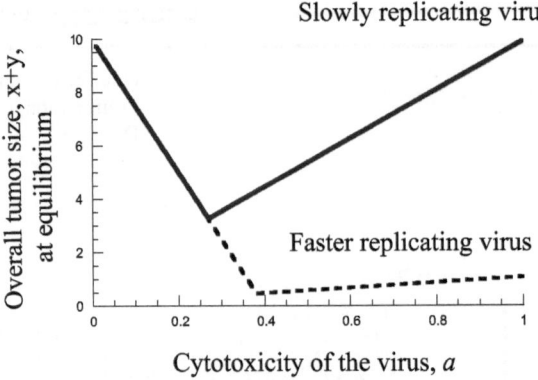

Fig. 12.2 Dependence of overall tumor load on the cytotoxicity of the virus. There is an optimal cytotxocity at which tumor load is smallest. The faster the rate of virus replication, the higher the optimal level of cytotoxocity, and the smaller the minimum tumor load. Parameters were chosen as follows: $\omega = 10$; $r = 0.2$; $s = 0.2$; $d = 0.1$; for fast viral replication, $\beta = 1$; for slow viral replication $\beta = 0.1$

In this case, two types of outcomes are possible. The virus can either attain 100% prevalence in the tumor cell population (i.e. all tumor cells are infected), or it may only infect a fraction of the tumor cells (i.e. both uninfected and infected tumor cells are observed). Hundred percent virus prevalence is described by equilibrium E1:

$$x^{(1)} = 0, \quad y^{(1)} = \omega(s - a)/s.$$

Coexistence of infected and uninfected tumor cells is described by equilibrium E2:

$$x^{(2)} = \frac{\beta\omega\,(a - s) + ar - sd}{\beta\,(\beta\omega + r - s)}, \quad y^{(2)} = \frac{\beta\omega\,(r - d) + sd - ra}{\beta\,(\beta\omega + r - s)}.$$

The virus infects all tumor cells (equilibrium E1) if

$$a < \frac{s(d + \beta\omega)}{r + \beta\omega}.$$

Otherwise, Equilibrium E2 is observed.

With this result in mind, how does viral cytotoxicity influence the size of the overall tumor? The tumor size is defined as the sum of infected and uninfected tumor cells, $x + y$, at equilibrium. Viral cytotoxicity has an opposing influence on tumor load depending on which equilibrium is attained (Fig. 12.2). If all tumor cells are infected, then

$$x + y = \frac{\omega(s - a)}{s}.$$

An increase in viral cytotoxicity results in a reduction in tumor load (Fig. 12.2). On the other hand, if not all tumor cells are infected, then

$$x + y = \frac{\omega(r - s + a - d)}{\beta\omega + r - s}.$$

Now, an increase in the viral cytotoxicity increases tumor load (Fig. 12.2). The reason is that increased rates of tumor cell killing reduce the amount of viral spread. This in turn increases the tumor load.

Hence, there is an optimal cytotoxicity, a_{opt}, at which the tumor size reaches a minimum. This optimum is the degree of cytotoxicity at which the system jumps from the equilibrium describing 100 % virus prevalence to the equilibrium where uninfected tumor cells are also present (Fig. 12.2a). The optimal viral cytotoxicity is thus given by

$$a_{opt} = \frac{s(d + \beta\omega)}{r + \beta\omega}.$$

At this optimal cytotoxicity the tumor size is reduced maximally and is given by

$$[x + y]_{min} = \frac{\omega(r - d)}{r + \beta\omega}.$$

There are a number of points worth noting about this result. The minimum tumor size this therapy regime can achieve is most strongly determined by the replication rate of the virus, β (Fig. 12.2). The higher the replication rate of the virus, the smaller the minimum size of the tumor. In order to achieve this minimum, the viral cytotoxicity must be around its optimum value. A major determinant of the optimal viral cytotoxicity is the rate of growth of uninfected and infected tumor cells (r and s respectively).

If the infected tumor cells grow at a significantly slower rate relative to uninfected cells ($s \ll r$), the optimal cytotoxicity is low (Fig. 12.3a). In the extreme case where the virus abolishes the ability of the tumor cell to divide, a non-cytotoxic virus is required to achieve optimal treatment results. More cytotoxic viruses result in tumor persistence (Fig. 12.3a).

On the other hand, if the growth rate of infected tumor cells is not significantly lower than that of uninfected tumor cells, an intermediate level of virus induced cell death is required to achieve minimum tumor size (Fig. 12.3b). If viral cytotoxicity is too weak, the tumor persists. However, if the viral cytotoxicity is too high, the tumor also persists because infected cells die too fast for the virus to spread efficiently (Fig. 12.3b). In general, the faster the replication rate of the virus, the higher the optimal level of cytotoxicity.

12.3 Effect of Virus-Specific CTL

As explained in Chap. 3, cytotoxic T lymphocytes or CTL or CD8+ T cells are a branch of the specific or adaptive immune response. These cells are the main immune mediators responsible for killing cells that are either infected with a virus or are aberrant due to tumor formation. They proliferate in response to viral or tumor proteins (antigens), and fight the unwanted cells. Therefore, they play an important role in the current context and will be included into the basic mathematical models of oncolytic virus dynamics. The current section explores the effect of anti-viral CTL responses.

We expand the above model to include a population of virus-specific CTL, z_v. The CTL recognize viral antigen on infected tumor cells. Upon antigenic encounter, the CTL proliferate with a rate c_v, proportional to the product yz_v. The term $p_v yz_v$ stands for the killing of the infected tumor cells by the CTL. In the absence of antigenic stimulation, the CTL die with a rate b. The model is given by the following set of differential equations [7]:

Fig. 12.3 Simulation of virus therapy in the absence of immunity (**a**) The growth rate of infected tumor cells is significantly slower than that of uninfected tumor cells. Parameters were chosen as follows: $\omega = 10; r = 0.5; s = 0; \beta = 1; d = 0.1.$ $a = 0.1$ for the non-cytotoxic virus, and $a = 0.5$ for the more cytotoxic virus. (**b**) The growth rate of infected tumor cells is not significantly reduced relative to that of uninfected cells. Parameters were chosen as follows: $\omega = 10; r = 0.5; s = 0; \beta = 1; d = 0.1. a = 0.2$ for the weakly cytotoxic virus, $a = 0.55$ for intermediate cytotoxicity, and $a = 3$ for strong cytotoxicity

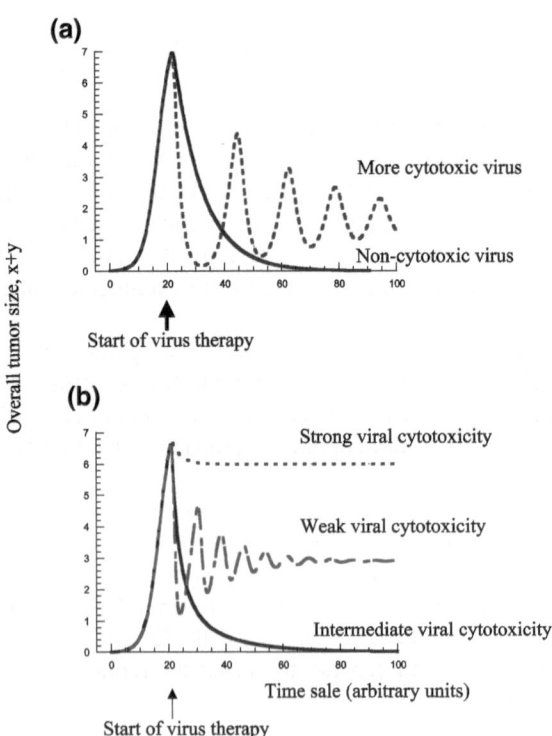

$$\dot{x} = rx\left(1 - \frac{x+y}{\omega}\right) - dx - \beta xy, \qquad (12.3)$$

$$\dot{y} = \beta xy + sy\left(1 - \frac{x+y}{\omega}\right) - ay - p_v y z_v, \qquad (12.4)$$

$$\dot{z}_v = c_v y z_v - b z_v. \qquad (12.5)$$

First, we define the conditions under which an anti-viral CTL response is established. This condition is different depending on whether the virus attains 100% prevalence in the tumor cell population in the absence of the CTL. The strength of the CTL response, or CTL responsiveness, is denoted by c_v. If the virus has attained 100% prevalence in the absence of CTL, the CTL become established if

$$c_v > \frac{bs}{\omega(s-a)}.$$

On the other hand, if the virus is not 100% prevalent in the tumor cell population in the absence of CTL, the CTL invade if

$$c_v > \frac{b\beta(\beta\omega + r - s)}{r(\beta\omega - a) - d(\beta\omega - s)}.$$

In the presence of the CTL, we again observe two basic equilibria: either the 100% virus prevalence in the tumor cell population, or the coexistence of infected and uninfected tumor cells. Hundred percent virus prevalence in the tumor cell population is described by equilibrium E1:

$$x^{(1)} = 0, \quad y^{(1)} = \frac{b}{c_v}, \quad z_v^{(1)} = \frac{\omega c_v(s-a) - sb}{p_v \omega c_v}.$$

Coexistence of infected and uninfected cells is described by equilibrium E2:

$$x^{(2)} = \frac{r(\omega c_v - b) - \omega(c_v d + b\beta)}{rc_v}, \quad y^{(2)} = \frac{b}{c_v},$$

$$z_v^{(2)} = \frac{\beta\omega(rc_v - b\beta - c_v d) - c_v(ar - sd) - b\beta(r - s)}{p_v c_v r}.$$

How do the CTL influence the outcome of treatment? We distinguish between two scenarios:

(i) If the virus has established 100% prevalence in the tumor cell population in the absence of the CTL response, the presence of CTL can both be beneficial and detrimental to the patient (Fig. 12.4): the virus can remain 100% prevalent in the tumor in the presence of CTL. In this case, overall tumor size is given by $x + y = b/c_v$. At this equilibrium, an increase in the CTL responsiveness against the virus decreases the tumor size (Fig. 12.4). On the other hand, if the

CTL responsiveness crosses a threshold given by

$$c_v > \frac{b(\beta\omega + r)}{\omega(r - d)},$$

the virus does not maintain 100% prevalence in the tumor cell population, and the overall tumor size is given by

$$x + y = \frac{\omega(c_v(r - d) - b\beta)}{c_v r}.$$

In this case, an increase in the CTL responsiveness to the virus increases tumor load and is detrimental to the patient (Fig. 12.4). This is because the CTL response kills the virus faster than it can spread. Hence, the optimal CTL responsiveness is given by

$$c_{opt} = \frac{b(\beta\omega + r)}{\omega(r - d)}.$$

At this optimal CTL responsiveness, the tumor size is reduced maximally and is given by

$$[x + y]_{min} = \frac{\omega(r - d)}{r + \beta\omega}.$$

The faster the replication rate of the virus, the higher the optimal CTL responsiveness, and the lower the minimum size of the tumor that can be attained by therapy (Fig. 12.4). Note that the minimum tumor size that can be achieved is the same as in the previous case, where viral cytotoxicity alone was responsible for reducing the tumor. The effect of the CTL response is to modulate the overall death rate of infected cells with the aim of pushing it towards its optimum

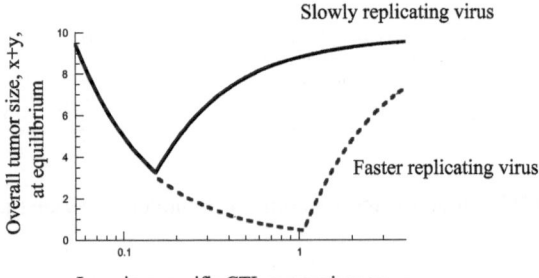

Fig. 12.4 Dependence of overall tumor load on the strength of the virus-specific CTL response. There is an optimal CTL responsiveness at which tumor load is smallest. The faster the rate of virus replication, the higher the optimal strength of the CTL response, and the smaller the minimum tumor load. Parameters were chosen as follows: $\omega = 10$; $r = 0.5$; $s = 0.5$; $d = 0.1$; $b = 0.1$; $p = 1$; $a = 0.2$. For fast viral replication, $\beta = 1$; for slow viral replication $\beta = 0.1$

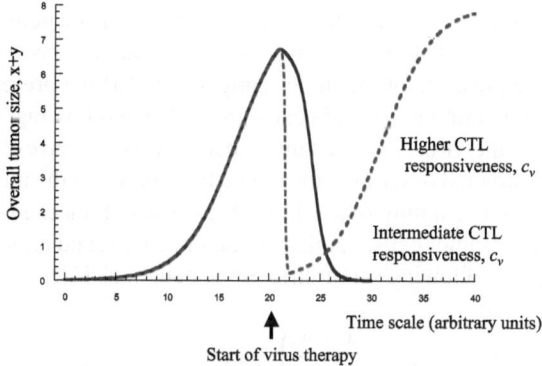

Fig. 12.5 Simulation of virus therapy in the presence of virus-specific lytic CTL. An intermediate CTL responsiveness results in tumor eradication, while a stronger CTL response results in tumor persistence. Parameters were chosen as follows: $\omega = 10$; $r = 0.5$; $s = 0.5$; $\beta = 0.1$; $a = 0.2$; $p = 1$; $b = 0.1$; $b = 0.1$; $d = 0.1$. The intermediate CTL responsiveness is characterized by $c_v = 0.2625$, while the stronger CTL response is characterized by $c_v = 2$

value. Figure 12.5 shows a simulation of therapy where an intermediate CTL responsiveness results in tumor remission, while a stronger CTL response can result in failure of therapy because virus spread is inhibited.

(ii) If the virus is not 100 % prevalent already in the absence of the CTL response, a CTL-mediated increase in the death rate of infected cells can only be detrimental to the patient since it increases tumor load by inhibiting the virus. The system converges to an equilibrium tumor size described by

$$x + y = \frac{\omega(c_v(r - d) - d\beta)}{c_v r}.$$

12.4 Virus Infection and the Induction of Tumor-Specific CTL

The above section explored how virus infection and the virus-specific CTL response can influence tumor load. However, virus infection might not only induce a CTL response specific for viral antigen displayed on the surface of the tumor cells. In addition, active virus replication could induce a CTL response specific for tumor antigens [2–5]. The reason is that virus replication could result in the release of substances and signals alerting and stimulating the immune system. This could be induced by tumor antigens being released and taken up by professional antigen presenting cells (APC), and/or by other signals released from the infected tumor cells. This is known as the danger signal hypothesis in immunology [4, 8]. Normal tumor growth is thought not to evoke such signals, whereas the presence of viruses can evoke danger signals. Here, such a tumor specific CTL response is included in the model. It is assumed that the responsiveness of the tumor-specific CTL requires two signals: (i) the presence of the tumor antigen, and (ii) the presence of infected tumor cells providing immuno-stimulatory signals . In the following, the interactions between the tumor, the virus, and the tumor-specific CTL are investigated.

A model is constructed describing the interactions between the tumor population, the virus population, and a tumor-specific CTL response. It takes into account three variables: uninfected tumor cells, x, infected tumor cells, y, and tumor specific CTL, z_T. It is given by the following set of differential equations [7]:

$$\dot{x} = rx\left(1 - \frac{x+y}{\omega}\right) - dx - \beta xy - p_T x z_T$$

$$\dot{y} = \beta xy + sy\left(1 - \frac{x+y}{\omega}\right) - ay - p_T y z_T$$

$$\dot{z}_T = c_T y z_T (x+y) - b z_T$$

The basic interactions between viral replication and tumor growth are identical to the models described above. The tumor-specific CTL expand in response to tumor antigen, which is displayed both on uninfected and infected cells $(x+y)$, at a rate c_T. However, in accord with the danger signal hypothesis, it is assumed that the tumor-specific CTL response only has the potential to expand in the presence of the virus, y. In the model virus load correlates with the ability of the tumor-specific response to expand, since high levels of viral replication result in stronger stimulatory signals. The tumor-specific CTL kill both uninfected and infected tumor cells at a rate p_T, proportionally to the product $y z_T$.

If the virus has reached 100 % prevalence in the absence of CTL, the tumor-specific CTL response becomes established if

$$c_T > \frac{bs^2}{(\omega(a-s))^2}.$$

If infected and uninfected tumor cells coexist in the absence of CTL, the tumor specific CTL response becomes established if

$$c_T > \frac{b\beta(s-r-\beta\omega)^2}{\omega(\beta\omega(r-d)-ra+sd)(r-s+a-d)}.$$

In the presence of the tumor-specific CTL, the virus can again attain 100 % prevalence in the tumor cell population, or we may observe the coexistence of infected and uninfected tumor cells. Hundred percent prevalence in the tumor population is described by equilibrium E1:

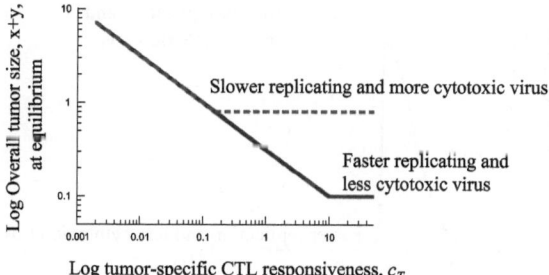

Fig. 12.6 Dependence of overall tumor load on the strength of the tumor-specific CTL response. The higher the strength of the tumor-specific CTL, the lower tumor load. If the strength of the tumor-specific CTL crosses a threshold, tumor load becomes independent of CTL parameters. The faster the rate of virus replication and the smaller the degree of viral cytotoxicity, the further the overall tumor load can be reduced. Parameters were chosen as follows: $\omega = 10$; $r = 0.5$; $s = 0.5$; $d = 0.1$; $b = 0.1$. The fast replicating and weakly cytotoxic virus is characterized by $\beta = 1$ and $a = 0.2$. The slower replicating and more cytotoxic virus is characterized by $\beta = 0.5$ and $a = 0.5$

$$x^{(1)} = 0, \quad y^{(1)} = \sqrt{\frac{b}{c_T}}, \quad z_T^{(1)} = \frac{\omega(s-a) - sy^{(1)}}{p_T \omega}.$$

Coexistence of infected and uninfected tumor cells is described by equilibrium E2, which is slightly cumbersome and we do not reproduce it here, please see [7].

We investigate how the responsiveness of the tumor-specific CTL, c_T, influences the size of the tumor, $x + y$. The presence of the tumor specific CTL can have the following effects. If the virus achieves 100 % prevalence in the tumor cell population, then

$$x + y = \sqrt{b/c_T}.$$

Thus, an increase in the responsiveness of the tumor-specific CTL results in a decrease in tumor load (Fig. 12.6). If

$$c_T > \frac{b(\beta\omega + r - s)^2}{(\omega(r - s + a - d))^2},$$

the virus is not 100 % prevalent in the tumor cell population. This switch is thus promoted by a high responsiveness of the tumor-specific CTL relative to the replication rate of the virus (Fig. 12.6). In this case, the size of the tumor is given by

$$x + y = \frac{\omega(r - s + a - d)}{\beta\omega + r - s}.$$

This is the minimum tumor size that can be achieved. Thus, if the CTL responsiveness against the tumor lies above a threshold, tumor load reaches its minimum (Fig. 12.6).

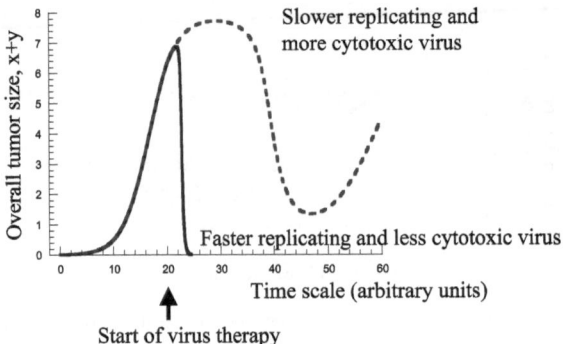

Fig. 12.7 Simulation of virus therapy with a tumor-specific CTL response. If the virus replicates at a fast rate and is weakly cytotoxic, the level of immune-stimulatory signals is high. Hence the tumor-specific response is strong and drives the tumor extinct. Parameters were chosen as follows: $\omega = 10$; $r = 0.5$; $s = 0.5$; $d = 0.1$; $b = 0.1$; $c_T = 0.2$. The fast replicating and weakly cytotoxic virus is characterized by $\beta = 0.5$ and $a = 0.2$. The slower replicating and more cytotoxic virus is characterized by $\beta = 0.1$ and $a = 0.6$

Note that it also becomes independent of the strength of the CTL. Hence, a CTL responsiveness that lies above this threshold is not detrimental to the patient. In this situation, tumor size is determined by the replication rate and the cytotoxicity of the virus (Fig. 12.6). The higher the replication rate of the virus and the lower the degree of viral cytotoxicity, the smaller the tumor. The reason is that fast viral replication and low cytotoxicity result in higher virus load which in turn results in stronger signals to induce the tumor-specific CTL. Figure 12.7 shows a simulation of treatment underscoring this result.

12.5 Interactions Between Virus- and Tumor-Specific CTL

In this section, the two types of CTL responses studied above are brought together. That is, both the virus- and the tumor specific CTL responses are taken into consideration. The model is explained schematically in Fig. 12.1 and given by the following set of differential equations [7].

$$\dot{x} = rx\left(1 - \frac{x+y}{\omega}\right) - dx - \beta xy - p_T x z_T$$

$$\dot{y} = \beta xy + sy\left(1 - \frac{x+y}{\omega}\right) - ay - p_v y z_v - p_T y z_T$$

$$\dot{z}_v = c_v y z_v - b z_v$$

$$\dot{z}_T = c_T y z_T (x+y) - b z_T$$

In this model the virus- and the tumor specific CTL responses are in competition with each other, because both can reduce tumor load and hence the strength of the stimulus required to induce CTL proliferation. In the following these competition dynamics are examined.

If the virus has reached 100% prevalence in the tumor cell population in the absence of CTL, then virus- and tumor specific CTL cannot coexist. If $c_v > (c_T b)^{1/2}$, then the virus-specific CTL response is established. On the other hand, if $c_v < (c_T b)^{1/2}$, then the tumor-specific CTL response becomes established.

If both infected and uninfected tumor cells are present in the absence of CTL, the situation is more complicated. Now, three outcomes are possible. Either the virus-specific response becomes established, or the tumor-specific response becomes established, or both responses can coexist. The virus-specific response persists if

$$c_v > \omega c_T \frac{r - s + a - d}{\beta \omega + r - s}.$$

The tumor-specific response persists if

$$c_T > \frac{c_v^2 r}{\omega (c_v(r - d) - b\beta)}.$$

Coexistence of both CTL responses is only observed if both of these conditions are fulfilled. This outcome is described by the following equilibrium expressions:

$$x^{(1)} = \frac{c_v^2 - bc_T}{c_v c_T}, \quad y^{(1)} = \frac{b}{c_v},$$

$$z_T^{(1)} = \frac{1}{p_T}\left[r\left(1 - \frac{x^{(1)} + y^{(1)}}{\omega}\right) - d - \beta y^{(1)}\right],$$

$$z_v^{(1)} = \frac{1}{p_v}\left[\beta x^{(1)} + s\left(1 - \frac{x^{(1)} + y^{(1)}}{\omega}\right) - a - p_T z_T^{(1)}\right]$$

If both responses coexist, then the size of the tumor is given by $x + y = c_v/c_T$. Thus, a strong tumor-specific response, c_T, reduces tumor load. On the other hand, a strong virus-specific response, c_v, increases tumor load. The reason is that a strong virus-specific response results in low virus load and therefore in low stimulatory signals promoting the induction of tumor-specific immunity. Note that this last statement only applies to the parameter region where both types of CTL responses co-exist.

12.6 Summary

The above discussion has shown that the outcome of therapy depends on a complex balance between host and viral parameters. An important variable is the death rate of infected tumor cells. In order to achieve maximum reduction of the tumor, the death rate of the infected cells must be around its optimum, defined by the mathematical models. If the death rate of infected cells lies around its optimum, a fast replication rate of the virus and a slow growth rate of the tumor increase the chances of tumor eradication. The death rate of infected tumor cells can be influenced by a variety of factors: (*i*) Viral cytotoxicity alone kills tumor cells. *(ii)* A CTL response against the virus contributes to killing infected tumor cells. *(iii)* The virus helps eliciting a tumor-specific CTL response following the release of immuno-stimulatory signals. This highlights the complexity that characterizes the correlates of successful virus therapies in settings that go beyond the basic interactions among viruses and target cells.

Problems

12.1. Numerical Project
CTL responses can not only kill virus-infected cells, but also secrete factors that reduce the rate of viral replication. This can be expressed by a variation of the infection term in Eqs. (12.3–12.5), given by $\beta xy/(qz_v + 1)$. Using analytical and numerical methods, explore the properties of the model under this assumption. What is the effect of the virus-specific CTL response?

12.2. Research Project
Find out about the danger signal hypothesis, and how infections or tissue injuries can potentially aid in cancer remission through induction of danger signals.

References

1. Wodarz, D., Nowak, M.A., Bangham, C.R.: The dynamics of htlv-i and the ctl response. Immunol. Today **20**(5), 220–227 (1999)
2. Melcher, A., Todryk, S., Hardwick, N., Ford, M., Jacobson, M., Vile, R.G.: Tumor immunogenicity is determined by the mechanism of cell death via induction of heat shock protein expression. Nat. Med. **4**(5), 581–587 (1998)
3. Kono, H., Rock, K.L.: How dying cells alert the immune system to danger. Nat. Rev. Immunol. **8**(4), 279–289 (2008)
4. Fuchs, E.J., Matzinger, P.: Is cancer dangerous to the immune system? Semin. Immunol. **8**(5), 271–280 (1996)
5. Matzinger, P.: An innate sense of danger. Semin. Immunol. **10**(5), 399–415 (1998)
6. Matzinger, P.: The danger model: a renewed sense of self. Science **296**(5566), 301–305 (2002)
7. Wodarz, D.: Viruses as antitumor weapons: defining conditions for tumor remission. Cancer Res. **61**(8), 3501–3507 (2001)
8. Gallucci, S., Matzinger, P.: Danger signals: sos to the immune system. Curr. Opin. Immunol. **13**(1), 114–119 (2001)

Chapter 13
Axiomatic Approaches to Oncolytic Virus Modeling

Abstract The previous models described oncolytic virus dynamics in the context of specific ordinary differential equations. However, the same biological assumptions can be expressed with alternative mathematical expressions, and the arbitrary formulations that are chosen can sometimes determine the behavior of the model. It is therefore desirable to formulate a model such that its properties do not depend on arbitrary mathematical terms that are used to describe biological processes. This can be achieved through the formulation of axiomatic models, where biological processes are described by general functions that are subject to reasonable constraints. This chapter formulates an axiomatic model of oncolytic virus dynamics and identifies different classes of models that share common properties. The different model classes and their properties are discussed, focusing on the conditions required for virus-mediated tumor control.

Keywords Axiomatic modeling · Tumor growth law · Infection terms · Equilibrium analysis · Phase portrait · Stability · Exponential growth · Linear growth · Surface growth · Ordinary differential equations · Differentiability · Generalized expansions

13.1 Introduction

The previous chapters looked at specific ODE models describing oncolytic virus dynamics. Chapter 11 considered the most basic model, and Chap. 12 elaborated on this by introducing further assumptions into the models. In both cases, the models are formulated in terms of specific mathematical expressions that describe various biological processes such as cell proliferation, cell death, and infection. In most cases, however, these mathematical expressions are rather arbitrary and several alternatives could describe the same process equally well. The "biologically true" mathematical form that should be used is not known due to lack of sufficient information. A good example is the infection term. In the models described in the previous chapters, it is given by *infection rate constant* × *number of susceptible cells* × *number of infected cells*, assuming perfect mixing of cells and viruses. While this type of infection term has been used extensively in the literature on virus dynamics [1, 2] and also in the context of infection models on the epidemiological level [3], the "correct" expression remains uncertain. Alternatives have also been used, assuming saturation in the number of infected and /or uninfected cells, such as frequency dependent infection terms [4]. Similar variants have also been implemented in models of oncolytic virus replication [5, 6]. This collection of papers also suggests that model properties can crucially depend on the exact mathematical terms chosen, which complicates the interpretation and biological usefulness of predictions. This chapter describes a general, axiomatic modeling approach that is designed to address these concerns. Instead of analyzing a specific model with arbitrary mathematical expressions, we consider a class of ODE models where biological processes are described by general functions that are subject to known biological constraints. We demonstrate how this approach can identify classes of models with different biological properties.

13.2 The Modeling Framework

Instead of considering specific ODE models, as done in previous chapters, we will take a generalized approach and consider a class of models. The general modeling framework used in our study is as follows. We take into account two populations: uninfected tumor cells, x; and infected tumor cells, y. The population of free viruses is not modeled explicitly. Because the turnover of free viruses is much faster than that of infected cells, we simply assume that the free virus population is in a quasi-steady state and proportional to the number of infected cells. The basic model is given as follows:

$$\dot{x} = xF(x, y) - \beta yG(x, y), \qquad (13.1)$$

$$\dot{y} = \beta yG(x, y) - ay. \qquad (13.2)$$

The function F describes the growth properties of the uninfected tumor cells, x, and the function G describes the rate at which tumor cells become infected by the virus. These functions are unknown and can potentially take a variety of forms, which will be discussed below. The coefficient β in front of the infection term represents the infectivity of the virus. Finally, virus-infected cells die with a rate ay.

This class of models is characterized by the existence of equilibria, the number and nature of which depends on the tumor growth term F and the infection term G. In the most general sense, the equilibria of the system are defined by the following two equations:

$$x F(x, y) = ay, \tag{13.3}$$

$$G(x, y) = \frac{a}{\beta}. \tag{13.4}$$

We will explore the equilibria and their properties depending on the tumor growth term F, and the infection term G.

The term F reflects the growth properties of an uninfected tumor. It comprises both division and death rates. The simplest assumption that can be made about the term F is that growth is exponential (or, more precisely, the division and death happen according to an exponential law, and the division rate is higher than the death rate). While this can be true during early stages of tumor growth, tumor growth certainly deviates from an exponential pattern at larger sizes for a variety of reasons, for example, space or nutrient limitations. Therefore, more complicated tumor growth terms involving some form of saturation must be considered [7]. In this respect, we can distinguish between two basic scenarios: First, while the rate of tumor growth saturates and slows down at higher tumor sizes, the tumor has the potential to keep growing towards infinity. Growth would stop once the tumor has reached a lethal size. Second, it can be assumed that growth not only slows down, but comes to a halt as the tumor size reaches a critical level, which can be called the carrying capacity of the tumor. This could happen when the division rate equals the death rate of the cells.

Regarding the infection term, $\beta y G(x, y)$, the assumption used most often in mathematical models is that it is directly proportional to the number of infected and uninfected cells, which means that $G(x, y) \sim x$ [4, 8]. This, however, assumes mass action or perfect mixing of populations, which is unrealistic, especially in the context of tumors. Instead, virus spread is likely to be slower, limited by spatial constraints. Since the virus released from one infected cell cannot reach all susceptible tumor cells in the population, the infection rate must be a saturating function of the number of susceptible tumor cells. Similarly, not all infected cells present in the population will be able to contribute to the generation of newly infected cells, for example if they are spatially separated from susceptible cells.

In the Methodology section, we list all the mathematical requirements that are imposed on the functions F and G. These requirements are motivated by the biological knowledge. Here we list the most important properties:

- For small values of x and y, the growth of cells in close to exponential ($F \approx 1$), and the spread of infection is also close to exponential.
- The function F cannot be an increasing function of x or y.
- The growth term, $\beta y G(x, y)$ increases with x and y.
- The growth *rate*, $\beta G(x, y)/x$, cannot increase with x and y.

In the sections following Methodology, we will define different classes of infection terms that have biologically reasonable characteristics, and investigate how they influence the properties of the model. These are based in part on mathematical work done in the context of infectious disease epidemiology [4, 8]. Subsequently, we will examine how changing the tumor growth term influences the model predictions.

13.3 Methodology

13.3.1 Axiomatic Model Construction

We assume that the cancer growth term satisfies the following biological requirements:

1. The obvious requirement of positivity: $F(x, y) \geq 0$ for all $x, y \geq 0$, and continuity.
2. At the beginning, the growth is exponential: $\lim_{z \to 0} F(z) = 1.$[1]
3. The growth slows down as the number of cells increases: $dF(z)/dz \leq 0$.
4. A symmetry requirement: $F(x, y) = F(x + y)$: the growth is controlled by infected and uninfected cells equally; note however that this requirement is not essential for our analysis.

We assume that the virus spread term $(\beta y G(x, y))$ satisfies the following biological requirements:

1. The obvious requirement of positivity: $G(x, y) \geq 0$ for all $x, y \geq 0$, and continuity.
2. For small values of x and y, the growth should be exponential to reflect perfect mixing:
$$\lim_{x,y \to 0} G(x, y)/x = 1.$$

3. The growth term must monotonically increase with x and y:

$$\frac{\partial(yG(x, y))}{\partial x} \geq 0, \quad \frac{\partial(yG(x, y))}{\partial y} \geq 0.$$

4. The growth *rate*, $G(x, y)/x$, must slow down with x and y:

$$\frac{\partial(G(x, y)/x)}{\partial x} \leq 0, \quad \frac{\partial(G(x, y)/x)}{\partial y} \leq 0.$$

5. The growth has to be saturated in both x and y, such that

$$\lim_{x \to \infty} yG(x, y) = H_x(y), \quad 0 < H_x(y) < \infty,$$

where $H_x(y)$ is a function of y independent of x. Similarly, with y.
6. For large values of x, the growth term cannot be positive in the limit of small y, that is,

---

[1] This requirement fixes the scaling of the time-variable. In general, if the initial growth-rate $\lim_{z \to 0} F(z) = r$, we scale time $t' = tr$, and also use $a' = a/r$ and $\beta' = \beta/r$. The primes are dropped for convenience.

$$\lim_{y \to 0} H_x(y) = 0.$$

Similarly, with x.

7. For large values of x and y the spread cannot stop completely:

$$\lim_{x,y \to \infty} yG(x, y) > 0.$$

Note that this expression could be infinite.

The analyses below rely on these properties.

13.3.2 The Number of Equilibria

The fixed points of the virus-cancer system (Eqs. 13.1 and 13.2) are given by $(0, 0)$ and all the solutions of the equations

$$xF(x + y) = ay, \tag{13.5}$$

$$G(x, y) = \frac{a}{\beta}. \tag{13.6}$$

The trivial point $(0, 0)$ has eigenvalues $F(0)$ and $-a$ and is thus a saddle. The number of solutions of Eqs. (13.5 and 13.6) depends on the particular shapes of the functions F and G. In order to find the nontrivial equilibria, we solve Eq. (13.5) to find $y(x)$, and then substitute it into Eq. (13.6). The equilibria are thus defined by the roots of equation

$$\beta G(x, y(x)) = a. \tag{13.7}$$

From Eq. (13.5) we can see that $y(0) = 0$. We know from assumption (2) on the function G that $G(0, 0) = 0$. The next step is to study the limiting behavior of $G(x, y(x))$ for large values of x. For that we need to know the behavior of $y(x)$ for large x. We have from Eq. (13.5):

$$\lim_{x \to \infty} y(x) = \lim_{x \to \infty} xF(x)/a.$$

There are three cases. (i) For a linear type growth, we have $\lim_{x \to \infty} xF(x) = c_0$, a nonzero constant. In this case, $\lim_{x \to \infty} y(x) = c_0/a$, with $0 < c_0 < \infty$. (ii) For any growth F which is superlinear but slower than exponential, we have $\lim_{x \to \infty} y(x) = \infty$, but $\lim_{x \to \infty} y/x = \lim_{x \to \infty} F(x + y)/a = 0$, that is, y increases slower than x. (iii) Finally, for exponential growth, $F = 1$ and $y(x) = x/a$, such that $y(x) \sim x$ for large values of x.

From the biological assumptions on the function $G(x, y)$ listed above, it follows that for any of the possible dependencies $y(x)$, the function $G(x, y(x))$ approaches a finite limiting value as $x \to \infty$, and this value can be zero or nonzero. We use the exponential case $F = 1$ and $y_{exp}(x) = x/a$ to separate all functions G into two classes. If $\lim_{x \to \infty} G(x, x/a) = 0$, then we will consider the virus spread to be *slow*, and if $\lim_{x \to \infty} G(x, x/a) = G_{exp}^{\infty} > 0$, with $G_{exp}^{\infty} < \infty$, we will regard this as *fast* spread. Note that for all laws of cancer growth slower than exponential, we have $G(x, y(x)) \geq G(x, y_{exp}(x))$. This is because $y(x) \leq y_{exp}(x)$, and G is a decreasing function of y.

13.3.3 Fast Virus Spread

In this case, the function $G_{exp}(x) \equiv G(x, y_{exp}(x))$ is either a monotonically increasing function (Fig. 13.3a), or it can attain one or more local extrema before converging to its nonzero horizontal asymptote.

For a monotonically increasing G_{exp}, low values of β correspond to zero roots in Eq. (13.6), which means that the cancer growth will continue indefinitely (Fig. 13.3a). As β crosses a critical value defined by a/G_{exp}^{∞}, there is one root. The value of x at this root drops as β increases (this is due to the convergence of G_{exp} to an asymptote, G_{exp}^{∞}). For large values of β, the value of x at the intersection tends to zero.

If the function G_{exp} has an absolute maximum at point x_{exp}, then an initial increase of β above a/c_{max} with $c_{max} = G_{exp}(x_{exp}, y(x_{exp}))$ results in the appearance of two roots. Additional local extrema will result in an appearance and disappearance of pairs of roots. However, as β increases through the second threshold given by a/c_2, only one (the lowest) root remains. The value of c_2 is given by the lower of the values $\{G_{exp}^{\infty}, c_{min}\}$, where c_{min} is the value at the lowest local minimum, if it exists.

In both cases, for sufficiently large values of β, there will be only one root in Eq. (13.6). Introducing other cancer growth laws can increase the limiting value of G thus decreasing the value of β_c. In the case of a monotonically increasing G_{exp}, there will be no qualitative change. If G_{exp} is one- or multiple-humped, the hump(s) may disappear. Whether this qualitative change happens depends on the relative size of the two spacial scales involved. The first scale is defined by the location of the maxima of G_{exp} and is related to the virus spread scale, s_v. The second scale is given by the size, s_t, at which cancer growth law starts to deviate from exponential. Once $s_t \sim s_v$, the limiting value of G becomes sufficiently large such that the "hump" disappears.

It is useful to investigate the value of x at the equilibrium as a function of β, for different values of s_t. Suppose that the graph of $G(x, x/a)$ is a monotonically growing function of x which approaches a limiting value, G_{exp}^{∞}. Suppose that the cancer growth slows down around the scales near s_t. So near $x \sim s_t$, the function $G(x, y(x))$ deviates from the horizontal asymptote, G_{exp}^{∞}, and starts growing toward a different, and higher horizontal asymptote, which we will call $G_1 > G_{exp}^{\infty}$ (see Fig. 13.1 for a particular example). The phase diagram as β increases is as follows. For $\beta < a/G_1$, there are no roots. As β crosses the first threshold, a/G_1, one root appears. The value of x at this equilibrium decays rapidly from infinity to values

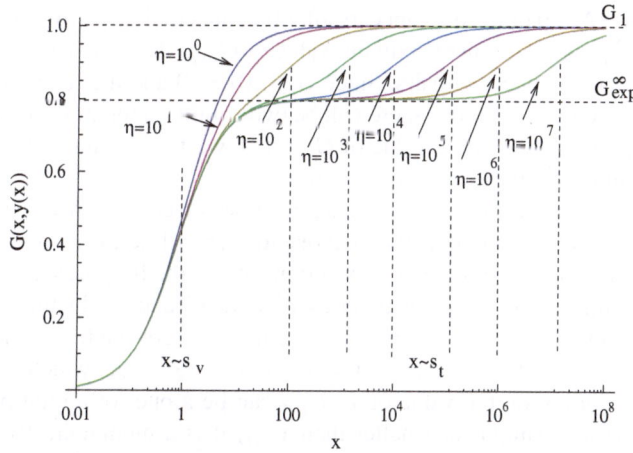

Fig. 13.1 The dependence of the equilibrium on β and s_t. We use the model with $F = \eta/(\eta+x+y)$ (thus, s_t is defined by η) and $G = x/(x + y + \varepsilon)$. The function $G(x, y(x))$ is plotted versus x for different values of η. The dashed vertical lines indicate the scales of interest: the leftmost such line corresponds to $x \sim s_v$, and the rest of the lines to $x \sim s_t$ for different values of η. The other parameters are: $a = 4$, $\varepsilon = 1$

around s_t, as β grows (because of the fact that G_1 is a horizontal asymptote). Then as β grows through its second threshold, a/G_{exp}^∞, the value of x at the equilibrium drops from s_t to values of order 1. The second transition is sharp if the following is satisfied: $s_t \gg x_1$, where x_1 is the value of x such that $|G(x_1, y(x_1)) - G_{exp}^\infty| = |G_1 - G_{exp}^\infty|$. In other words, x_1 is the value of x where the function $G(x, y(x))$ comes near its first horizontal asymptote ("near" means that it is at least as close to G_{exp}^∞, as G_{exp}^∞ is to G_1). If $s_t \gg x_1$, then the function G has a significant interval in x where it comes near the value G_{exp}^∞, before it deviates from it to start growing toward G_1. This guarantees a threshold effect.

We conclude that for all cancer growth laws and for all functions G corresponding to a fast virus spread, increasing β beyond a threshold leads to the existence of only one equilibrium, whose value correlates negatively with the viral replication rate, β. For large enough s_t, there is a "threshold" effect, such that the size at equilibrium decreases very sharply as β approaches a defined value.

13.3.4 Slow Virus Spread

In this case, the function $G_{exp}(x) \equiv G(x, y_{exp}(x))$ is a one- or a multiple-humped function, which approaches zero for large x (Fig. 13.3b).

In the case of an exponential growth, the bifurcation diagram looks as follows. As before, small values of β correspond to no equilibria (zero roots in Eq. (13.6)).

As we increase β, a pair of roots appears after the threshold given by a/c_{max}, where $c_{max} = G(x_{exp}, y(x_{exp}))$, the maximum value of G_{exp}. As β increases, other roots may appear and disappear in pairs. Since $G_{exp}(0) = 0$ and the function G_{exp} has zero as its horizontal asymptote, there will be two equilibria for all values of β larger than a/c_{min}, where c_{min} is the value of G at its lowest local minimum in the case that such a minimum exists; $c_{min} = c_{max}$ otherwise.

Let us next consider how non-exponential laws of cancer growth modify this picture. In some cases, a slower-than-exponential growth term, F, will lead to the horizontal asymptote of $G(x, y(x))$ becoming nonzero. In general, whether this happens depends on the functional forms of both G and F. In the case of linear growth, $y(x) \equiv y_{lin}(x)$ converges to a nonzero constant, c_1, and we have $\lim_{x \to \infty} G_{lin}(x) = \lim_{x \to \infty} G(x, c_1) = G_x(c_1) = c_2 < \infty$, which is a nonzero constant. Depending on the value of s_t, G_{lin} can be a one- or a multiple-humped function, or (for s_t similar or smaller than x_{exp}) it is a monotonically increasing function of x. In either of these cases, there exists a finite value of β such that for all values of β larger than this value, there is only one root in Eq. (13.6). For growths faster than linear but slower than exponential, we have $y \to \infty$ as x grows, but $y = o(x)$, i.e., it grows slower than x. In some cases the function G will retain a zero asymptote (e.g. in the case where $G = x/(x + 1)/(y + 1)$ and a surface growth law for F, see Table 13.1). In other cases it will acquire a nonzero limit (e.g., with $G = x/(x + 1 + \sqrt{x}(y + 1))$ and a surface growth law for F). In the latter case we can say that the surface cancer growth is sufficiently slow to warrant successful treatment given the particular mode of viral spread.

13.3.5 Finite Tumor Size

Here, we consider a tumor growth term which becomes zero for a finite value of $x + y$. The growth starts off exponential ($F(0) = 1$) and at some size, s_t, it slows down (we do not exclude the possibility that $s_t \sim 1$, that is, the growth becomes slower than exponential right away). Then there exists another characteristic size, $W \gg s_t$ such that the growth slows down further and stops. We define W such that $F(W) = 0$. Note that if $s_t \sim W$ then there is no need to introduce the two scales, s_t and W. Therefore, the assumption $s_t \ll W$ must hold.

The previous analysis holds on the scales intermediate between s_t and W, such that $s_t \ll x \ll W$. In particular, for values $x \ll W$, the shape of the curve $G(x, y(x))$ is similar to that obtained for the corresponding unlimited growth. As x grows far beyond s_t and approaches W, the function G approaches $G(W, 0)$. If, for the unbounded growth, the limiting value of the G function is c_2, we have in general $G(W, 0) \geq c_2$, and the curve G takes an upward turn in the vicinity of $x = W$. This means that Eq. (13.6) acquires an additional root corresponding to the cancer growing to its carrying capacity, W. In the systems with unrestricted growth this was equivalent to an unlimited growth of the cell population.

It is useful to note the following: in systems with a limited size, the function $G(x, y(x))$ is always bounded away from zero. Therefore, strictly speaking, we can always find a threshold value β_t such that for $\beta > \beta_t$, only one root is present. However, if $W \gg s_v$, such values of β are very large compared to β_c, and in most cases are probably not achievable.

13.3.6 Stability

Let us suppose that (x_0, y_0) with $x_0 \geq 0$ and $y_0 \geq 0$ is a solution of system (13.5 and 13.6), and consider the stability of the corresponding equilibrium (for stability theory of ODEs please see, e.g., [9]). The Jacobian of the system can be written as a 2×2 matrix, $\{m_{ij}\}$, with

$$m_{11} = F + x_0 F' - \beta y_0 G_x, \quad m_{12} = x_0 F' - \beta(G + y_0 G_y),$$
$$m_{21} = \beta y_0 G_x, \quad m_{22} = \beta y_0 G_y.$$

Here the functions F and G are assumed to be differentiable (see below for remarks regarding violations of this assumption), and their derivatives are evaluated at the point (x_0, y_0): $G_x = \partial G/\partial x|_{x=x_0, y=y_0}$, and similarly with G_y and F'. The equilibrium is stable if the following two conditions hold:

$$m_{11} + m_{22} < 0, \tag{13.8}$$
$$m_{11}m_{22} - m_{21}m_{12} \geq 0. \tag{13.9}$$

13.3.6.1 Saddle Points

Condition (13.9) is equivalent to the positivity of the derivative of G in the direction defined by the implicit relation $ya = xF(x+y)$, Eq. (13.5). Differentiating it, we get: $ady = Fdx + xF'(dx+dy)$. The directional derivative is equal to $(G_x dx + G_y dy) = [G_x(a - F'x_0) + G_y(F + x_0 F')]/(a - F'x_0)$. The denominator is positive, so this expression has the same sign as the left-hand side of condition (13.9).

The equilibria are defined by the roots of Eq. (13.7). Since $G(0, 0) = 0$, all the odd roots of Eq. (13.7) will correspond to a positive, and the even ones to a negative slope of the left hand side of Eq. (13.7). This means that all even equilibria are saddles. This is because in such cases the directional derivative is negative, condition (13.9) is violated, and therefore there are two real eigenvalues of opposite signs. On the other hand, an odd root can be either a sink, a source, or a spiral (stable or unstable). This is because for such a root, condition (13.9) is always satisfied, so that we could either have complex eigenvalues, or real roots of the same sign (positive or negative).

In the presence of a saddle, an infinite outcome (corresponding to an unchecked cancer growth) is possible. For large values of x, we have

$$\dot{x} = x F^{\infty} - \beta y G^{\infty}(y), \tag{13.10}$$
$$\dot{y} = y(\beta G^{\infty}(y) - a), \tag{13.11}$$

where $\lim_{x \to \infty} G(x, y) = G^{\infty}(y)$ and $\lim_{x \to \infty} F(x, y) = F^{\infty}$. The growth of y becomes negative as y increases if $\lim_{y \to \infty} G^{\infty}(y) = 0$, which suggests that y settles to a finite value which makes the right-hand side of Eq. (13.11) zero, such that the outcome $(\infty, const)$ is observed. If $\lim_{y \to \infty} G^{\infty}(y) = const > 0$, then for large enough values of β we can have an outcome of the form (∞, ∞).

13.3.6.2 Properties of the Internal Equilibrium

Let us first show that for large values of β, there will be an equilibrium, (x_0, y_0), such that $\lim_{\beta \to \infty} x_0 = 0$ and $\lim_{\beta \to \infty} y_0 = 0$. We call this equilibrium the "internal equilibrium." Its existence follows from Eq. (13.7) and the properties of the function G. We know that $y(0) = 0$, and also that $G(0, 0) = 0$. It is also clear that there is an interval of x, $[0, \xi]$, where G is a growing function. Therefore, by continuity, for all $\beta \ge a/G(\xi, y(\xi))$, there will be a solution of Eq. (13.7). From monotonicity of the function G, the value of x at the intersection with a/β decays with β. From Eq. (13.5) it follows that there is an interval of x, $[0, \xi_1]$, where y is a growing function of x. Therefore, we conclude that for large enough β, there is an equilibrium whose x and y values decay with β and approach 0 in the limit $\beta \to \infty$.

Let us evaluate the left-hand sides of inequalities (13.8) and (13.9) for small values of x_0 and y_0. First, we express β from Eq. (13.6): $\beta = a/G(x_0, y_0)$. Then we approximate the curve $y(x)$ by its Taylor series for small values of x_0:

$$y_0 = F x_0/a + (a + F)F'(x_0/a)^2 + (a + F)((F')^2 + 1/2(a + F)F'')$$
$$(x_0/a)^3 + O[(x_0/a)^4], \tag{13.12}$$

where the function F and its derivatives are evaluated at 0. This expression follows from expanding both sides of Eq. (13.5) in Taylor series in terms of x_0 and y_0, solving for y_0 and using a Taylor expansion of this expression. Now, let us multiply the left-hand side of inequality (13.8) by $G(x_0, y_0)$, and use expression (13.12). Expanding in terms of small x_0, we obtain:

$$G(x_0, y_0)(m_{11} + m_{22}) = (F'G_x + G_{xy} - G_{xx}/2)x_0^2$$
$$+ \frac{1}{a}\left((a + 1)F''G_x + (a + 2)F'G_{xy} + G_{xyy}\right.$$
$$+ \frac{1}{2}((a - 1)G_{xxy} - F'G_{xx}) - \frac{1}{3}aG_{xxx}\left.\right)x_0^3$$
$$+ O([x_0]^4). \tag{13.13}$$

Here, the functions F and G and their derivatives[2] are evaluated at zero. To derive the above expression we also used the fact that the function G and its y-derivatives are equal to zero if $x = 0$, and $F'(0) = 1$.

Next, we evaluate the left-hand side of inequality (13.9) in the same manner:

$$G(x_0, y_0)(m_{11}m_{22} - m_{21}m_{12}) = aG_x x_0 + (aF'G_x + 2G_{xy} + aG_{xx})x_0^2 + O([x_0]^3).$$

We can see that the expression above is always positive, so condition (13.9) is satisfied for large enough values of β. This means that the two eigenvalues (if real) have the same sign. Condition (13.8) however is not necessarily satisfied, as follows from expression (13.13). The expansion can be positive or negative, depending on the particular properties of the functions F and G.

Next, we would like to investigate whether the eigenvalues are real or complex. For the eigenvalues to have an imaginary part, the following condition has to be satisfied:

$$(m_{11} - m_{22}) + 4m_{12}m_{21} < 0. \tag{13.14}$$

Performing a Taylor expansion of the above expression for small values of x_0 and y_0 at internal equilibrium, we obtain:

$$G(x_0, y_0)((m_{11} - m_{22}) + 4m_{12}m_{21}) = -4aG_x^2 x_0^2 - 2G_x$$
$$(2aF'G_x + 6G_{xy} + 3aG_{xx})x_0^3 + O([x_0]^4).$$

We can see that this quantity is always negative. Therefore, we conclude that the internal equilibrium has complex eigenvalues for sufficiently large values of β.

13.4 Different Classes of Infection Terms and Their Properties

Let us consider two different classes of viral growth, see Fig. 13.2a,b.

Tumor-virus systems belonging to class I are characterized by the following property: if the number of uninfected tumor cells is high relative to the number of infected cells, virus growth does not slow down as the number of infected cells rises. Virus growth is exponential. Biologically, this can be interpreted as virus replication in a non-solid tumor where cells mix relatively freely. In other words, infected cells are not clustered together in a mass but are interspersed among uninfected cells. This is shown schematically at the top of Fig. 13.2a (the white circles represent uninfected cells, and the black circles—infected cells). In this case, if the number of uninfected cells is relatively large, then every infected cell is likely to be surrounded by uninfected cells to which the virus can be passed on. Alternatively, a similar picture can be achieved by a very high motility of the virus. In either case, all infected cells

---

[2] Here we assume that the functions F and G are differentiable at zero. Non-differentiable functions are handled similarly by using generalized expansions.

(a)   **(b)**

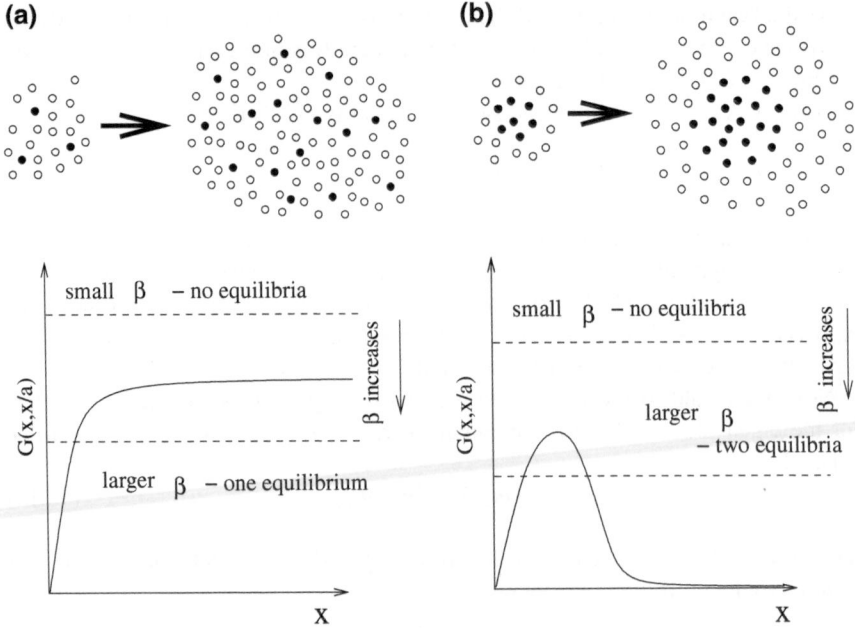

Fig. 13.2 Two classes of virus growth captured by the mathematical models. (**a**) According to class I or fast virus growth, virus growth is exponential as long as the number of uninfected cells is significantly larger than the number of infected cells. This can correspond to a high degree of mixing between infected and uninfected cells. As the virus population grows, the number of cells that contribute to virus spread remains constant because most infected cells will have an uninfected cell in their vicinity. (**b**) According to class II or slow virus growth, virus growth slows down and saturates as the virus population increases in size, even if the number of uninfected cells is relatively large. This can correspond to spatial clustering of the infected cells. Only infected cells at the surface have uninfected cells in their neighborhood and can thus contribute to virus transmission. As the number of infected cells rises, the number of "active" cells that can contribute to virus transmission declines

contribute to viral spread and growth is exponential. We call this "fast virus spread." On the other hand, with tumor-virus systems that belong to class II, the virus growth rate decreases as the number of infected cells rises, even if the number of uninfected cells is very large. The biological interpretation is that infected tumor cells are clustered together, Fig. 13.2b. This can occur in solid tumors, which typically show a high degree of spatial arrangement. In this case, as the number of infected tumor cells increases, most infected cells will be surrounded by other infected cells and not by uninfected cells. Hence, they cannot pass on the virus and cannot contribute to virus spread. Only cells at the periphery of the infected cell mass have uninfected cells in the neighborhood and can contribute to new infection events. We refer to this model of infection as "slow virus spread."

Next let us connect this classification with the mathematical model, and in particular, with the infection term, $\beta y G(x, y)$. The function $G(x, y)$ is related to the proportion of the total population of the infected cells which participates in the infection process. It is plotted in Fig. 13.2 as a function of the number of tumor cells, x, and we examine the shape of these plots. Let us take a closer look at the schematic at the top of Fig. 13.2b. Because of the geometrical arrangement of the cells in this case, only the infected cells on the surface of the black core will be able to infect other cells (it is 6 out of the 7 cells in the smaller colony presented). Now, let us increase the system size, such that the number of infected and uninfected cells grows in the same proportion. Again, only the infected cells close to the surface of the infected core will participate in the infection process. However, now the proportion of the surface cells is much smaller (11 out of 20 cells). As the size of the system increases, the proportion of such "active" cells (that is, cells capable of infecting other cells) decreases. This is what is depicted in the graph in Fig. 13.2b, where the function $G(x, y)$ declines following the peak.[3] Next, we take a look at the cell arrangement at the top of Fig. 13.2a. Here, the populations are well-mixed, and as the system grows, a constant fraction of infected cells will be able to infect new cells. This is reflected in the corresponding graph of $G(x, x/a)$, which reaches an asymptote and does not decline. In Table 13.1a we list several examples of fast and slow growth laws.

In general, we can prove that the two scenarios above are the only possible outcomes, given the biological requirements imposed on the function G. As x increases, this function increases, and can either approach zero or a nonzero level. If it approaches a nonzero level, this does not necessarily need to occur via a monotonic approach to the asymptote. It is possible that the function G first increases, peaks,

Table 13.1 Examples of different cancer growth terms and virus spread terms

(a) Virus spread terms, $G(x, y)$

$G(x, y)$	Law of virus spread
$(\varepsilon + 1)x/(x + y + \varepsilon)$	Fast (frequency dependent)
$\dfrac{x}{\sqrt{x+y+\varepsilon_1}(\sqrt{x}+\sqrt{y}+\varepsilon_2)}$	Fast
$\dfrac{(\varepsilon_1+1)(\varepsilon_2+1)x}{(x+\varepsilon_1)(y+\varepsilon_2)}$	Slow
$\dfrac{x}{(\sqrt{xy}+\varepsilon_1)(\sqrt{x}+\sqrt{y}+\varepsilon_2)}$	Slow
$\dfrac{x}{\sqrt{x}(y+c)+x+\varepsilon}$	Slow

(b) Tumor growth terms, $F(x + y)$

$F(x + y)$	Growth law
1	Exponential
$\eta/(\eta + x + y)$	Linear
$(\eta/(\eta + x + y))^{-1/2}$	Surface growth in 2D
$(\eta/(\eta + x + y))^{-1/3}$	Surface growth in 3D
$1 - (x + y)/W$	Logistic
$\log \dfrac{W+\eta}{x+y+\eta} \left(\log \dfrac{W+\eta}{\eta}\right)^{-1}$	Gompertzian

---

[3] For very small system sizes, the proportion of cells participating in infection is formally zero because of the lack of uninfected cells, therefore the graph of the function G starts at zero, reaches a peak, and then declines for high values of x.

and then converges to a nonzero asymptote. For intermediate values of x the function G may have a more complicated structure than that shown in Fig. 13.2, but in the absence of any biological evidence of that it is a safe bet to assume the simplest shape with a minimal number of local extrema.

How does the shape of G help us draw meaningful conclusions about the behavior of the biological system? It turns out that the function G is essential in determining the number and the stability properties of the equilibria of the system, and thus it will help us reason about long-term predictions on the treatment outcome.

Equations (13.3 and 13.4) can be combined in a single equation

$$G(x, y(x)) = a/\beta, \tag{13.15}$$

where the function $y(x)$ is a relationship between the number of infected and uninfected cells at equilibrium as the total system size grows; it is obtained from Eq. (13.3) and depends on the exact rate of cancer growth, F. If the cancer growth is exponential ($F = 1$), we have $y(x) = x/a$, that is, at equilibrium, the infected cells comprise a fixed fraction of uninfected cells. Thus the function $G(x, x/a)$ depicted in Fig. 13.2 is just the left-hand side of the equation for the equilibria, Eq. (13.15). The right-hand side is represented by horizontal dashed lines, whose level decreases with the viral replication rate β. The number of intersections corresponds to the number of equilibria in the system.

We can see that the two graphs in Fig. 13.2 exhibit different numbers of equilibria. First we consider Fig. 13.2a, fast virus spread. In this case, the model always contains a parameter region in which exactly one equilibrium exists. If the viral replication rate, β, lies below a threshold ($\beta < \beta_c$) then no equilibrium exists. If the viral replication rate lies above that threshold, the following is observed. As shown in Fig. 13.2a exactly one equilibrium is found. In other cases, it is possible that there are two or more equilibria for intermediate viral replication rates. For example, if the function $G(x, x/a)$ first rises and achieves a maximum before descending to its horizontal asymptote, or if it goes through a number of local extrema before approaching a horizontal asymptote. The most important universal feature in all fast growth scenarios is that for sufficiently high values of β, there is exactly one equilibrium.

Next, consider Fig. 13.2b, slow virus spread. Again, for an equilibrium to exist, the viral replication rate needs to lie above the threshold $\beta > \beta_c$. If this is the case, the system is always characterized by the presence of not one, but two equilibria. Again, in some cases, it is possible that the intermediate values of β correspond to more than two equilibria.

The biological interpretation of this analysis is as follows. We saw that for both modes of infection, if the values of the viral replication rate β are small, no equilibria exist. This translates into an uncontrolled cancer growth. This is an intuitive result: for low viral replication rates, treatment is impossible. A less intuitive result is connected with the number of equilibria once β is above its threshold value.

The cancer-virus system displays a fundamentally different behavior depending on whether it is characterized by one or two equilibria. If there is only one equilibrium, then the dynamics will be governed by the properties of this equilibrium only. Because the number of tumor cells is relatively low at this equilibrium, this outcome corresponds to containment of the tumor by the virus. For convenience, we call this internal equilibrium E_I. On the other hand, the situation is more complicated if the system is characterized by two equilibria. The first equilibrium, at which the number of tumor cells is lower, is again the internal equilibrium, E_I, and can be interpreted as containment of the tumor by the virus. The second equilibrium can be shown to be an unstable saddle node equilibrium, call it E_S. The presence of the saddle equilibrium means that the dynamics are qualitatively different depending on the initial conditions. If the initial number of tumor cells is relatively low and close to the internal equilibrium, then the dynamics are governed by this internal equilibrium, E_I, leading to a degree of tumor control. If the initial number of tumor cells is higher and around or above the saddle node equilibrium E_S, then the number of tumor cells increases in an uncontrolled fashion. Hence, in this regime, uncontrolled cancer growth is always a possible outcome.

We conclude that our biologically defined modes of virus spread correspond to very different mathematical properties. Models of class I (fast virus spread) contain a parameter region (of high enough β) in which only a single equilibrium is observed. In this case, the model contains a parameter region in which uncontrolled cancer growth is impossible. Models of class II (slow spread) never have only one equilibrium and the saddle node equilibrium E_S is present whenever the internal equilibrium E_I exists. In this class of models, no matter how high β is, uncontrolled cancer growth is always a possibility.

13.5 Effect of the Tumor Growth Term

For the purposes of classification of the virus spread terms, we looked at the changes in G as the number of infected and uninfected cells grew in the same proportion. This led to a direct evaluation of the number of equilibria for exponential cancer growth ($F = 1$). While mathematically the simplest scenario, exponential growth is an unrealistic assumption, because the growth of cells is bound to saturate as the tumor grows. Our methods allow to study any reasonable cancer growth law in a very natural way.

Let us model a slowdown of the tumor growth rate as the number of tumor cells increases. This can be done in two different ways. On the one hand, we can assume that while tumor growth slows down, it never stops, such that the tumor can grow toward infinity over time. That is, there is no upper limit to the number of tumor cells; in practical terms, growth will stop when the organism dies. An example is what we call "surface growth," where only the cells around the surface of the tumor can give rise to viable daughter cells and can contribute to tumor spread. This can apply to solid tumors that have a high degree of spatial structure. Surface growth in 2D and 3D are listed in Table 13.1b. The parameter η determines the tumor size at which saturation comes into play. Another possibility that falls into this category is that the rate of tumor growth becomes linear as the number of tumor cells increases. In this case, tumor growth is even slower; we refer to it as "linear growth."

On the other hand, it is possible that there is a natural limit or carrying capacity W, that limits tumor growth [7]. Thus, we will assume that growth slows down and eventually stops as the number of tumor cells increases. This can occur in a variety of ways. Tumor growth can be exponential until the number of cells approaches carrying capacity and the rate of tumor cell growth becomes zero. For example, this can be described by the logistic growth term (see Table 13.1b) [7]. Alternatively, we can assume that tumor growth first saturates according to the surface growth or linear growth patterns described above, and only reaches the carrying capacity once the tumor has grown to a significantly larger number of cells. Another example of a growth with a carrying capacity is a Gompertzian-type growth [7], Table 13.1b.

As mentioned before, the term F reflects implicitly both division and death properties of uninfected tumor cells. For example, an exponential growth is characterized by a net expansion rate resulting from exponential division and death processes. The logistic growth is a consequence of saturation of the division rate while the (exponential) death rate remains constant. In fact, any process with a sub-exponential division rate and an exponential death will be characterized by a finite carrying capacity. On the other hand, an unlimited (but saturated) growth (such as surface growth) implicitly includes death which happens slower than exponentially. If we were to add an exponential death term to a surface growth, it would lead to a limited growth with a carrying capacity. Our framework includes all these and any other reasonable functional forms of cellular growth.

In the following, we will examine the effect of different types of tumor growth terms on the properties of the model. We will do this first in the context of the faster virus infection terms that belong to class I, and then in the context of the slower infection terms that belong to class II. Note that our analysis is quite general and the particular growth laws listed in Table 13.1 are merely an illustration; the results are not restricted to these particular growth laws.

13.5.1 Effect on Fast Virus Growth

With this class of virus infection term, we found that in the context of exponential tumor growth, $G(x,y(x))$ with $y(x) = x/a$ approaches a nonzero asymptote for large values of x (note that it can either rise monotonically to the asymptote, or first go through one or more local maxima before declining toward the asymptote). In either case, for any equilibrium to exist, the viral replication rate needs to lie above a threshold $\beta > \beta_c$, and there exists a parameter region (characterized by values of β greater than a threshold) in which only the internal equilibrium E_I is present. In this parameter region, tumor control is the only outcome.

Introducing saturated tumor growth (or changing the function F in any way) will lead to a different functional form of $y(x)$ in Eq. (13.15). A universal feature is that any tumor growth slower than straight exponential growth will lead to smaller

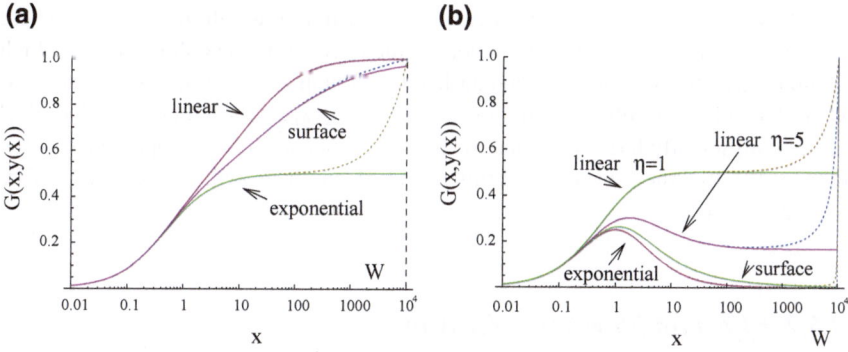

Fig. 13.3 The effect of a carrying capacity. The function $G(x,y(x))$ is plotted for two particular choices of the virus spread law and three different laws of cancer growth: exponential, surface growth and linear growth. (**a**) Fast virus spread, $G(x, y) = x/(x+y+1)$ and (**b**) slow virus spread, $G = x/(x+1)/(y+1)$. The solid lines correspond to the unlimited cancer growth; the dotted lines—to a growth up to a given size, W. The parameters are: $a = 1$, $\eta = 10$ and $W = 10^4$

values of $y(x)$ and thus to higher values of G. Therefore, as a result of tumor growth saturation, the asymptote becomes higher for slower tumor growth terms. This means that only the internal equilibrium E_I can exist, as with exponential growth. The only difference lies in the viral replication rate threshold beyond which this equilibrium can exist and beyond which tumor control is possible, see Fig. 13.3a. The slower the tumor growth, the lower the viral replication rate threshold required for virus-mediated control.

If we assume saturated but limited tumor growth (i.e., growth stops at carrying capacity W), then the picture is similar for the most part, with one difference. After the term $G(x,y(x))$ has approached the asymptote, the curve G takes an upward turn in the vicinity of $x = W$, i.e., when the number of cells approaches carrying capacity. This means that the model acquires an additional equilibrium, which corresponds to the cancer growing to its carrying capacity W. In the systems with unrestrictive growth, this was equivalent to unlimited growth of the cell population to infinitely large sizes. This is illustrated with the dotted line in Fig. 13.3a. We can see that for $x \ll W$, the curves for limited and unlimited growth laws look identical, and near the carrying capacity W they deviate.

So far, we have concentrated on the case where $G(x,y(x))$ increases monotonically toward an asymptote. Alternatively, the term $G(x,y(x))$ can rise to a peak and then decline toward a nonzero asymptote. In this case, including saturation into the tumor growth term $F(x, y)$ leads to similar consequences. However, the hump in the function can disappear, eliminating any parameter region in which both equilibria can exist. In other words, with slower tumor growth, there is no parameter region anymore in which the tumor can escape the effect of the virus and grow out of control. Whether this occurs or not depends on the relative size of the two spatial scales involved. The first scale is defined by the tumor size at which the virus infection function G saturates and peaks in the context of exponential growth; this is entirely dependent on the properties of the viral growth term. Let us call this scale s_v, where the subscript refers to "viral." The second scale is given by the colony size at which the tumor growth law starts to deviate from exponential; we will call this scale s_t (where the subscript refers to "tumor"). When $s_t \leq s_v$, the asymptotic value of G becomes sufficiently large such that the hump disappears. The disappearance of the hump makes treatment easier, and this occurs if tumor growth slows down before virus growth does.

13.5.2 Effect on Slow Virus Growth

Here, we assume slower virus growth terms that belong to class II. In the context of exponential growth, the function $G(x,y(x))$ first increases, and then declines toward zero. This means that if equilibria exist, both the internal equilibrium E_I and the saddle node equilibrium E_S are aways present. Consequently, the possibility always exists that the tumor can out-run the virus infection and grow uncontrolled. Taking into account saturated tumor growth has the following effect (Fig. 13.3b). (*i*) The function G can remain qualitatively the same; that is, it rises to a peak and then declines toward zero. *(ii)* Alternatively, the picture can change such that it does not decline toward zero, but toward a nonzero asymptote, while remaining a one-humped function. *(iii)* Finally, the picture can change further such that the function G increases monotonically toward an asymptote. Which outcome is observed depends on the exact nature of the functions F and G and also the relative size of the two spatial scales involved: the tumor size at which the virus infection term G saturates and peaks (s_v), and the size s_t at which the pattern of tumor growth starts to deviate from exponential. Lowering the value of s_t relative to s_v shifts the outcome from scenario (*i*) to *(iii)*. As the value of s_t becomes similar to the value of s_v, the model contains parameter regions in which only the internal equilibrium E_I exists and in which uncontrolled tumor growth is impossible. If $s_t \ll s_v$, then the hump in the function G disappears, and the saddle node equilibrium E_S is never present. In this case, virus-induced tumor control is the only outcome, and uncontrolled tumor growth cannot be observed. In biological terms, saturation of tumor growth at lower sizes promotes successful virus therapy. These arguments apply to all saturated tumor growth scenarios. With saturated but unlimited tumor growth, the function G approaches an asymptote for large tumor sizes x. For tumor growth that is limited by a carrying capacity W, the function G eventually deviates from the asymptote and rises again, indicating the presence of an equilibrium that describes tumor growth toward carrying capacity rather than toward infinity. Lowering the carrying capacity W has the same effect as lowering the parameter s_t that determines the tumor size at which growth starts to saturate: it shifts the outcome from scenario (*i*) to *(iii)*.

13.6 Summary of Model Properties

In summary, this analysis has provided the following insights. We examined two types of infection terms and found that they strongly influence the dynamics of oncolytic virus spread. In the first class of models, virus spread is fast. In this case, tumor control is always observed if the viral replication rate lies above a threshold. In these parameter regions, loss of tumor control is not observed. In the second class of models, virus spread is slow. In this situation, the model can be characterized by bistability. If the initial number of tumor cells lies below a threshold, tumor control is observed. If the initial number of tumor cells lies above this threshold, uncontrolled tumor growth is observed. If tumor growth only saturates at high numbers of tumor cells or not at all, then uncontrolled tumor growth is always possible in parameter regions in which tumor control is possible. If tumor growth saturates at lower levels, there are parameter regions in which only the tumor control outcome is observed and in which uncontrolled tumor growth is not possible. If tumor growth saturates at even lower levels, then the bistability and the dependence on initial conditions vanishes completely.

13.7 Properties of the Internal Equilibrium

The above analysis concentrated on the equilibria. By examining which equilibria exist under different conditions, we can obtain information about the ability of the virus to control the cancer, and about the possibility that the cancer grows despite the presence of the virus. If the dynamics are governed by the internal equilibrium E_I, then the virus keeps the tumor cell population at relatively low levels and prevents uncontrolled tumor growth. We have discussed the conditions under which this can be achieved and interpreted these conditions from a biological angle. If the virus does control the tumor, however, additional questions arise. The virus can either control a persisting tumor at low levels, or the virus can drive the tumor cell population extinct.

Because we are considering ordinary differential equations that describe the average behavior of the cell and virus populations, true extinction cannot occur in this model. The number of cells can, however, drop to very low levels. If the average number of cells is below one, we can assume that tumor extinction is a likely event. Therefore, if the number of tumor cells at equilibrium lies below one, we can say that the virus is likely to drive the tumor extinct. However, even if the equilibrium number of cells lies above one, the tumor cell population can still go extinct during oscillatory dynamics that can occur before the dynamics reach equilibrium. Therefore, we need to understand the properties of the internal equilibrium in more detail. We will examine this in the context of both fast and slow virus growth. We will only assume saturated tumor growth and not consider straight exponential tumor growth.

13.7.1 Fast Virus Growth

One of the most important parameters that influence the properties of the internal equilibrium is the replication rate of the virus β. In general, the faster the replication rate of the virus, the lower is the equilibrium number of tumor cells. Further, it can be shown that if the viral replication rate β crosses a threshold, the behavior near the equilibrium becomes oscillatory. Both promote the eradication of the cancer. In general, the internal equilibrium can either be stable or unstable, depending on the particular model under consideration as well as parameter values.

Let us first consider the case where the equilibrium is stable. We can distinguish between two parameter regions. Denote the size at which tumor growth slows down and deviates from exponential by s_t. In the first parameter region, the value of s_t is large compared to a value related to the virus scale, s_v. In this parameter region, we observe a viral replication rate threshold, at which the equilibrium number of tumor cells drops sharply from relatively high values to values of the order 1 (Fig. 13.4a). This replication rate threshold can be defined for individual models that belong to this class and defines the condition for cancer eradication. If the tumor size at equilibrium drops to small values (of the order of 1 cell), stochastic effects are very likely to lead to extinction. This is further supported by changes in the oscillatory approach to the equilibrium, which we have investigated in the context of individual models (Fig. 13.4b). At this viral replication rate threshold, the amplitude of the initial oscillations can increase sharply, as can the time it takes for the dynamics to approach the stable equilibrium (the real part of the eigenvalues of the Jacobian matrix rapidly approaches zero). Since pronounced oscillations reduce the number of tumor cells well below one, tumor eradication is the likely outcome. Note that this drastic change in the oscillatory pattern is not observed in all models that belong to this class. The sharp drop in the equilibrium value is, however, a universal feature of models that belong to this class.

Now assume the other parameter region in which the scale s_t is small. In this case, no such viral replication rate threshold exists. Instead, the equilibrium number of tumor cells declines gradually, inversely proportional to the viral replication rate β. Numerical simulation of individual models, however, indicates that the *minimum* number of tumor cells can decline exponentially with an increase in the viral replication rate, although this could not be proved in general. In other words, even though the equilibrium number of tumor cells does not experience any threshold behavior, the approach to equilibrium often involves population oscillations, and as a result of such oscillations, the population drops to low levels, which exponentially decrease with β. Taken together, these findings indicate that in the parameter regions where virus replication is fast enough such that there is an oscillatory approach to the equilibrium, tumor eradication is the likely outcome.

As mentioned above, it is also possible that the internal equilibrium E_I is unstable. In this case, we observe oscillations that diverge away from the equilibrium if the viral replication rate β is sufficiently fast. That is, the amplitude of the oscillations increases over time. This is likely to correlate with extinction of the tumor, especially

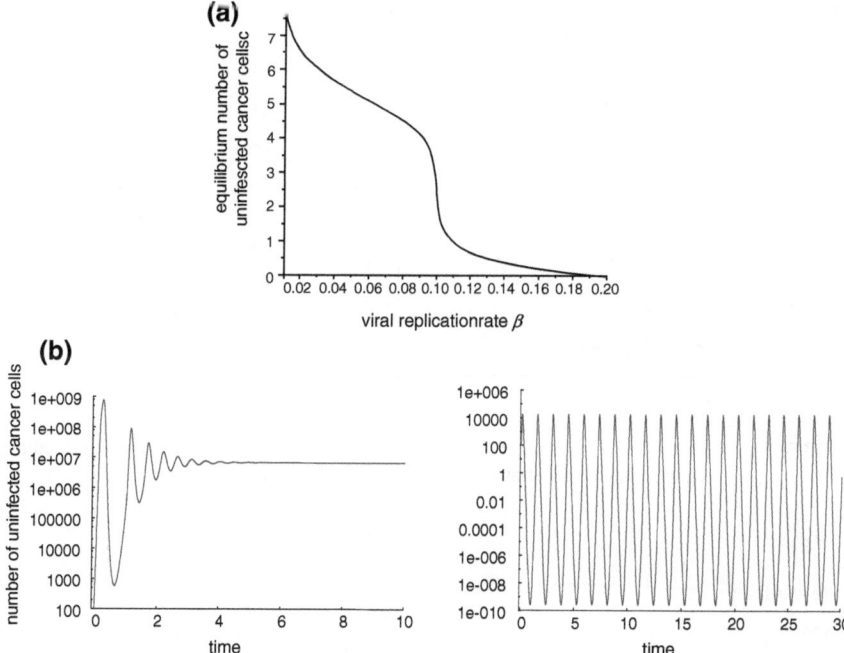

Fig. 13.4 (**a**) The equilibrium number of uninfected cancer cells as a function of the viral replication rate β for fast virus growth. There is a threshold viral replication rate at which the number of cancer cells drops sharply from relatively high values to values of the order of one. This can be considered a tumor extinction threshold. (**b**) Dynamics of the uninfected cancer cells if the viral replication rate lies below (*left*) and above (*right*) this threshold. If the viral replication rate lies below the threshold, limited oscillations are observed that dampen out quickly. If the viral replication rate lies above the threshold, extensive oscillations are observed that reduce the cancer cell population to very low levels, and that dampen out very slowly (dampening is not observed on timescale shown here). These plots were made by using a specific model from the fast virus growth category, that is $G = (\varepsilon + 1)x/(x + y + \varepsilon)$. Note that the transition in oscillations is not a universal feature of all models in this class. Parameters were chosen as follows: (**a**) $r_1 = 1$; $a = 0.1$; $\varepsilon = 10$; $\eta = 10^8$; $x_0 = 100$; $y_0 = 10$; For (**b**), $\beta = 0.07$ and $\beta = 0.13$, respectively

if the number of tumor cells at equilibrium is relatively low. This is because the oscillations will reduce the number of tumor cells well below the equilibrium value over time. Thus we conclude that for sufficiently large values of β, the cancer will be driven extinct by the virus through (convergent or divergent) oscillations.

13.7.2 Slow Virus Growth

In this case, the tumor size at the internal equilibrium is again negatively correlated with the viral replication rate β. Similarly to fast virus growth, the internal equilibrium

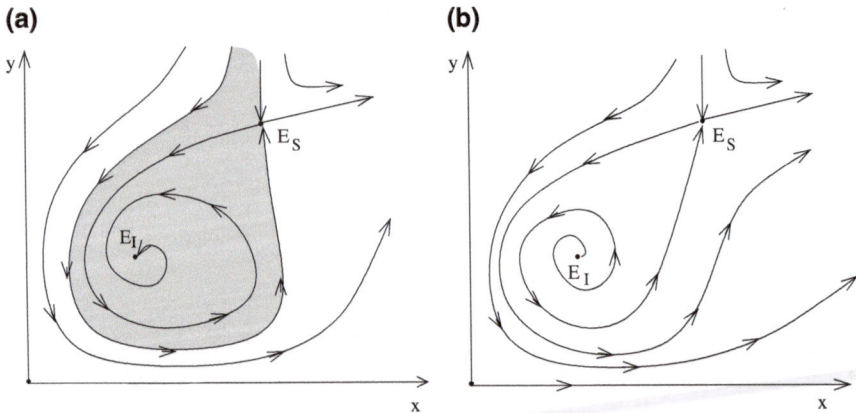

Fig. 13.5 The phase portrait for a system with a slow virus propagation term. (**a**) The intermediate equilibrium, E_I, is stable (the basin of attraction is shaded), (**b**) E_I is unstable

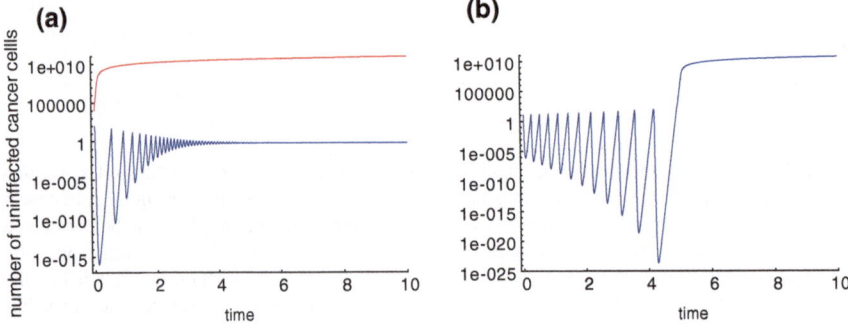

Fig. 13.6 Dynamics in fast virus growth models assuming that the internal equilibrium E_I is (**a**) stable and **b** unstable. (**a**) If the internal equilibrium is stable, then the dynamics can converge to this equilibrium via damped oscillation if the initial number of cancer cells is relatively low. On the other hand, if the initial number of cancer cells is relatively high, then uncontrolled cancer growth is observed. (**b**) If the internal equilibrium is unstable, then diverging oscillations are observed. Eventually, these diverging oscillations take the populations beyond the saddle node equilibrium, leading to unlimited cancer growth. Before that occurs, however, it is most likely that the cancer has been driven extinct in a stochastic setting because the diverging oscillations drive the tumor size to ever decreasing values. These plots were obtained from a specific model that belongs to the slow virus growth class, i.e. $G = \frac{(\varepsilon_1+1)(\varepsilon_2+1)x}{(x+\varepsilon_1)(y+\varepsilon_2)}$. Parameters were chosen as follows: **a** $r = 1$; $\beta = 0.8$; $a = 0.5$; $\varepsilon_1 = 20$; $\varepsilon_2 = 10$; $\eta = 10^8$; $x_0 = 100$ and $10,000$, respectively; $y_0 = 10$. For (**b**) $r = 1$; $\beta = 1$; $a = 0.5$; $\varepsilon_1 = 10$; $\varepsilon_2 = 11$; $\eta = 10^8$; $x_0 = 10$; $y_0 = 1$

can be stable or unstable depending on the individual model and on the parameter values. The dynamics will be discussed for both stable and unstable equilibria E_I.

If the equilibrium is stable, the approach is again oscillatory if the viral replication rate is sufficiently large (Fig. 13.5a and 13.6a). Numerical simulations of individual models indicate that the minimum tumor size during these oscillations can decline exponentially with the viral replication rate β, although again this could not be proven in a general setting. These results indicate that if oscillations are observed, it is likely that the cancer is eradicated by the virus (Fig. 13.6). Note that this assumes that the initial number of tumor cells is sufficiently small such that the population is in the region of attraction of the internal equilibrium. If this is not the case, the virus falls and unlimited cancer growth takes place because the long-term outcome depends on the initial conditions as discussed above (Fig. 13.6a).

On the other hand, if the internal equilibrium is unstable, then the following is observed (Fig. 13.5b and 13.6b). If the viral replication rate is fast enough, the populations show diverging oscillations away from the equilibrium (Fig. 13.5b), i.e., the amplitude of the oscillations increases over time. During these diverging oscillations, the minimum number of tumor cells declines over time. Hence, the tumor is likely to hit extinction.

To summarize, for slow virus growth oscillations around the internal equilibrium have the potential to drive the tumor cell population extinct. However, the bi-stability of this system causes problems since there is always the possibility that the populations can escape to large numbers, leading to uncontrolled tumor growth.

13.8   Summary

This chapter showed how an axiomatic modeling approach can be used to analyze oncolytic virus dynamics in the context of ordinary differential equations (ODEs). This approach does not apply specifically to oncolytic viruses, but can and should be used in the context of any models that aim to describe the dynamics of biological systems. The vast majority of ODE models of biological processes use specific equations that contain mathematically arbitrary expressions. The same process can be described equally well with alternative mathematical expressions, which could alter model properties and predictions. Therefore, any prediction based on such analysis may not be robust. The approach showcased here does not suffer from this drawback, and predictions derived with such a methodology are far more robust. Given certain known biological constraints, the axiomatic modeling approach allows us to understand the full set of behaviors that are possible.

Problems

13.1 For different virus growth terms in Table 13.1a, show if they are "fast" or "slow" by using the definitions of these terms.

13.2 Take $G(x, y) = \frac{x(1+\varepsilon)}{x+y+\varepsilon}$, and study the properties of the internal equilibrium under a general cancer growth term, $F(x + y)$, see [6].

13.3 Research prject

(a) Take $G(x, y) = \frac{x}{\sqrt{x+y+\varepsilon}(\sqrt{x}+\sqrt{y}+\sqrt{\varepsilon})}$, and study the properties of the internal equilibrium under a general cancer growth term, $F(x + y)$, see [6]. The function is not differentiable in y at $y = 0$, and a generalized expansion must be used.

(b) Modify the infection term slightly, such as $G(x, y) = \frac{x}{\sqrt{x+y+\varepsilon}(\sqrt{x+\delta_1}+\sqrt{y+\delta_2}+\varepsilon)}$.

Now the function is differentiable at 0, and conditions (13.8 and 13.9) can be used. Study the properties of the internal equilibrium.

13.4 Research prject

Consider $G(x, y) = \frac{x(1+\varepsilon)}{x+y+\varepsilon}$ and $F = 1 - (x + y)/W$. Study the singular behavior of the system as $\varepsilon \to 0$, see [6]. Find the equilibria, and study their stability properties as ε decays to zero.

13.5 Numerical prject

Take $G(x, y) = \frac{x(1+\varepsilon_1)(1+\varepsilon_2)}{(x+\varepsilon_1)(y+\varepsilon_2)}$ and study analytically and numerically the number and stability properties of internal equilibria as β increases. Take several different shapes for the function $F(x + y)$, such as exponential growth and limited growth, see [6].

References

1. Nowak, M.A., May, R.M.: Virus dynamics: Mathematical principles of immunology and virology. Oxford University Press, New York (2000)
2. Wodarz, D.: Killer Cell Dynamics: Mathematical and Computational Approaches to Immunology. Springer, New York (2006)
3. Anderson, R.M., May, R.M.: Infectious Diseases of Humans. Oxfors University Press, Oxofrd (1991)
4. McCallum, H., Barlow, N., Hone, J.: How should pathogen transmission be modelled? Trends Ecol. Evol. **16**(6), 295–300 (2001)
5. Wodarz, D., Komarova, N.: Towards predictive computational models of oncolytic virus therapy: basis for experimental validation and model selection. PLoS ONE **4**(1), e4271 (2009)
6. Komarova, N.L., Wodarz, D.: Ode models for oncolytic virus dynamics. J. Theor. Biol. **263**(4), 530–543 (2010)
7. Adam, J.A., Bellomo, N.: A survey of models for tumor-immune system dynamics. Birkhauser, Boston (1997)
8. Begon, M., Hazel, S.M., Baxby, D., Bown, K., Cavanagh, R., Chantrey, J., Jones, T., Bennett, M.: Transmission dynamics of a zoonotic pathogen within and between wildlife host species. Proc. Biol. Sci. **266**(1432), 1939–1945 (1999)
9. Arrowsmith, D.K., Place, C.: An Introduction to Dynamical Systems. Cambridge University Press, Cambridge (1990)

Chapter 14
Spatial Oncolytic Virus Dynamics

Abstract Tumors are often characterized by intricate spatial structures. Therefore, understanding the principles of virus spread in spatially structured tumor cell populations is of fundamental importance. This chapter describes a two-dimensional spatial model where cells and viruses can place their offspring only into the direct vicinity. This describes a cell culture situation where a virus spreads through a target cell population that is arranged in a two-dimensional monolayer. According to this model, infection can give rise to different types of spatial patterns. On the one hand, an expanding ring of infected cells can be observed that eventually eliminates the target cell population. On the other hand, a more diffuse growth pattern can occur where infected and uninfected cells mix without forming macroscopic patterns. This typically leads to failure to eliminate the target cell population. Conditions are defined under which the different outcomes are observed. Interestingly, nonspatial, ordinary differential equations that are applied to local neighborhoods of the system can predict the outcome of the entire spatial simulation. This can have implications for evaluating the ability of viruses to spread in a spatial setting by means of simple in vitro experiments that lack spatial structure.

Keywords Oncolytic viruses · Spatial models · Plague formation · Nearest neighbor · Cellular automaton · Agent-based modeling · Virus growth patterns · Stochastic simulations · Time-series · Extinction · Invasion · Local dynamics · Viral persistence · The ring structure · The disperse phenotype · In vitro experiments · Adenovirus

14.1 Introduction

Chapter 13 investigated the dynamics of oncolytic virus replication in the context of ordinary differential equations. This means that dynamics was deterministic and that perfect mixing of cells and viruses was assumed. The focus was on identifying conditions under which the virus can drive the tumor cell population extinct. However, most tumors are characterized by intricate spatial structure, and the effect of this

N. L. Komarova and D. Wodarz, *Targeted Cancer Treatment in Silico*,
Modeling and Simulation in Science, Engineering and Technology,
DOI: 10.1007/978-1-4614-8301-4_14, © Springer Science+Business Media New York 2014

on virus dynamics needs to be explored with alternative modeling approaches. This chapter reviews a spatially explicit agent-based model of oncolytic virus dynamics. It studies the interactions between the virus and its target cells in a two-dimensional setting where cells and viruses can only place offspring into the direct neighborhood. This model is also stochastic in nature, which is useful when examining the conditions for tumor extinction. Various layers of complexity can be incorporated into such models. This chapter concentrates on one of the simplest possible models and shows that even in this case, very complex dynamics can be observed. A detailed understanding of this forms the foundation for exploring more complex scenarios.

14.2 A Spatially Explicit Agent-based Model of Oncolytic Virus Dynamics

Tumors are characterized by intricate spatial structures that cannot be captured by ordinary differential equations. Other modeling methods have to be called upon in order to capture this complexity. However, the spatial complexity of tumors is enormous and not well understood, which can undermine the predictive power of such spatial models. As a first step to tackle this problem, we considered a relatively simple in vitro setting, in which target cells are arranged in a two-dimensional monolayer and in which viruses can only spread from the infected source cell to the nearest neighboring target cells [1]. This relatively simple experimental setting can be described by an agent-based model, and model predictions can be tested relatively easily by experiments. In the model, each cell is represented as an "agent" occupying a certain position on a grid, and interacting with other cells according to some (probabilistic) rules. Our modeling approach is spatial, that is, it takes into account the spatial distribution of the uninfected and infected cells. The model, based on [2], describes target cell-virus dynamics on a two-dimensional grid that contains $N \times N$ spots [1]. Each spot is either occupied by a cell (infected or uninfected), or it is empty. We model the development of the populations in discrete time. Given the state of the system at time t, a set of rules is applied to each spot, and this gives rise to the state of the system at time $t+1$. At each time step, the grid is randomly sampled N^2 times. If the chosen spot is occupied by an uninfected cell, it can die with a probability D, leaving the spot empty. Alternatively, the cell can reproduce with a probability R, and a destination spot is randomly chosen for the offspring from the set of eight nearest neighboring spots. If the destination spot is empty, the offspring is placed there, otherwise, no reproduction occurs. If the chosen spot contains an infected cell, it can die with a probability A, or attempt to transmit the virus with a probability B. A destination spot is chosen randomly from the eight nearest neighbors, and infection only proceeds when a susceptible cell is present. Infected cells are assumed not to reproduce. Typical oncolytic viruses lock the cell in the S-phase for replication, thus preventing further divisions [3].

14.3 Initial Virus Growth Patterns

In this section, we explore the initial virus growth patterns. As starting conditions, we assume that the grid is filled with uninfected cells and that a relatively small square of infected cells (30 × 30 spots) is placed in the middle of the grid (which overall contains 300 × 300 spots). The emerging growth pattern depends on parameters that influence the rate of virus spread, in particular the probability for an infected cell to die, A, and the probability for an infected cell to transmit the virus, B. The patterns that we observed are presented in Fig. 14.1.

Figure 14.1 shows both the spatial configurations of infected and uninfected cells, and their quantifications. Each row in the figure represents one case, characterized by a certain parameter combination. The left graph shows the spatial pattern. Green indicates uninfected cells, red infected cells, and gray empty spots. The middle graph shows the number of infected cells over time. The different lines represent 100 different instances of the simulation with the same parameter combination. The right graph shows the square root of the number of infected cells over time, again showing lines for 100 different runs. If this graph is linear, we observe quadratic or "surface" growth, see below.

Two basic types of initial growth are observed. The cases of Figs. 14.1a and b show the formation of a ring structure, where the infected cell population expands as a ring or wave that leaves no uninfected cell behind in its core. The two pictures differ in the death probability of infected cells. In Fig. 14.1b, the death probability of infected cells is relatively high, such that during the time frame of the simulation, the cells infected earlier die, forming a structure that looks like a hollow ring. In Fig. 14.1a, the probability for infected cells to die is relatively low such that during the time frame of the simulation a hollow ring has not yet formed and the infected cell population expands as a relatively solid mass.

It is possible to show that the total number of cells in scenarios of Figs. 14.1a, b is proportional to

$$\frac{e^{-At} - 1 + At}{A^2},$$

see [1]. For short time scales (or smaller death rates) the growth is quadratic in time, and for longer times scales (or larger death rates) it is linear in time. This is exactly what is observed. Figure 14.1a, characterized by smaller values of A, shows a growth law of the infected cell population that is close to quadratic. In Fig. 14.1b, where the death rate is larger, the infected cell population grows linearly once the hollow ring is present. Note that these two scenarios are identical in principle. In Fig. 14.3a, the death rate of infected cells is lower, thus requiring a longer period of time (and a larger grid) until a hollow ring is formed. Hence, the duration of the quadratic growth phase is relatively long, almost the entire duration of the simulation presented here. Eventually, however, it will transition to linear growth, as exhibited in case Fig. 14.3b, characterized by a faster death rate of infected cells. The higher the death rate of infected cells, the faster the ring is formed, and the faster the growth law changes from square to linear.

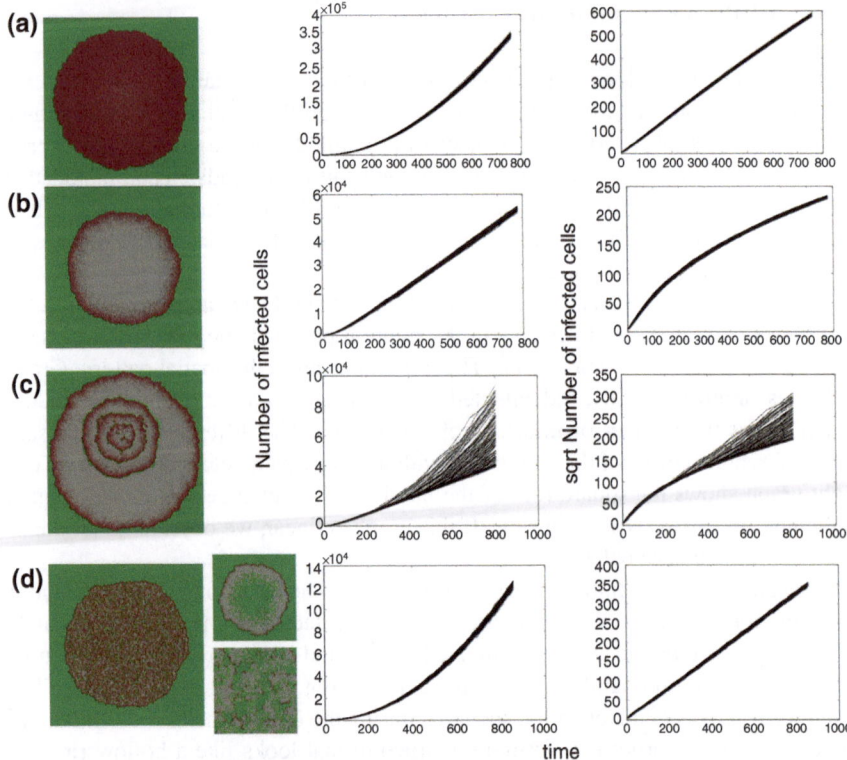

Fig. 14.1 Initial virus growth dynamics in the agent-based model. *green* indicates uninfected cells, *red* infected cells, and *grey* empty spots. The middle graph shows the number of infected cells over time. The different lines represent 100 different instances of the simulation with the same parameter combination. The *right* graph shows the square root of the number of infected cells over time, again showing lines for 100 different runs. Parameter values were chosen as follows. (**a**) $R = 0.5; D = 0;$ $B = 0.6; A = 0.601$. (**b**) $R = 0.5; D = 0; B = 0.6; A = 0.62$. (**c**) $R = 0.5; D = 0; B = 0.6;$ $A = 0.628$. (**d**) $R = 0.5; D = 0; B = 0.6; A = 0.7$. The small graphs in (**d**) are characterized by $R = 0.04$, leading to fewer target cells in the area of infection and thus to slower viral spread

Lowering the rate of virus spread (decreasing the value of B and increasing the value of A) gives rise to the patterns of a different nature.

In Fig. 14.1c, the wave of infection is less effective, and uninfected cells are left behind in the core of the expanding ring. When they grow and become infected by virus, a coupled expanding ring of uninfected and infected cells forms. This can occur repeatedly, giving rise to concentric rings. The persistence of cells in the core of the ring is probabilistic in nature, and that is reflected in the growth laws that are observed in multiple runs of the simulation. In cases where uninfected cells are not left behind inside the ring, the infected cell population grows linearly. When concentric rings do occur, the growth becomes quadratic.

Finally, no expanding ring structure is formed in Fig. 14.1d, because the viral spread kinetics are even slower. Instead, the area of virus growth is characterized by a mix of infected and uninfected cells that expands over time. We call this type of viral spread a "disperse pattern". In this case, quadratic growth of infected cells is observed. Note that if the viral spread kinetics are in the lower end of this spectrum, it is possible to observe a variation of this pattern, as shown in the inset of Fig. 14.1d: while the spreading infection leaves uninfected cells behind, the viral spread kinetics are too low to maintain significant numbers of infected cells throughout this area. Most of the infected cells will be at the outer edge of the infection due to a higher density of target cells. In this case, a relatively thin, ring-like structure can be formed, with a large area of uninfected cells remaining in its core. This pattern, however, is temporary. With time, one of two scenarios can be observed. A mixed pattern can be generated, characterized by a large number of uninfected cells and a low number of infected cells, because the virus eventually spreads to the remaining susceptible cells. Alternatively, there is a chance that the virus population spontaneously goes extinct due to the slow rate of spread. Long-term outcomes are discussed further below.

Note that the case of Fig. 14.1c lies between the hollow ring and the disperse pattern. Initially, a ring structure is formed, resulting first in quadratic, then in linear growth. However, on rare occasions uninfected cells remain in the wake of the expanding virus population, thus leading to concentric rings. While this does not happen in all instances of the simulations, when it does occur, the growth law transitions back from linear to quadratic.

14.4 Growth Patterns and the Extinction of Cells

Here, we explore the long-term dynamics, investigating how the above described patterns play out and correlate with the overall outcome if both the uninfected and infected cell population can expand in space. We seek to define conditions under which the virus can eliminate the target cell population in this system. All simulations are started with a small number of infected cells placed in a compact vicinity (a 5×5 neighborhood) into a larger space filled with uninfected cells (a 13×13 square), which is in turn embedded into an even larger "empty" space.

We go beyond the initial virus growth stage, and focus on time scales where the population of target cells experiences significant changes (grows in size in the absence of infection). The dependency of the system behavior on the parameters is investigated (Fig. 14.2), by running at least 10^4 instances of the simulation, where the $\log_{10}$ of all the parameters was varied between -4 and 4. For each parameter combination (that is, for different values of the $(R/B, A/B)$ pairs), we record the system behavior concentrating on the long-term outcome. These outcomes are presented by means of the following color code (Fig. 14.2): Blue indicates coexistence of virus and cells. Red and orange indicate extinction of the cells and thus the virus. Red is used if extinction occurs before the boundary of the system has been reached, while orange is used if extinction occurs after cells have reached the boundary of the system. Grey indicates extinction of the virus while cells persist.

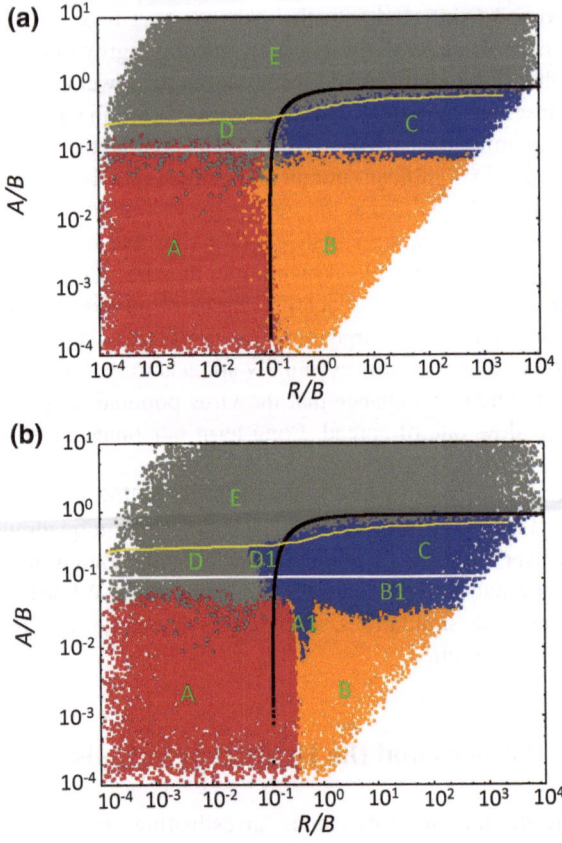

Fig. 14.2 Dependence of outcomes on parameters in the agent-based model for (**a**) a 30 × 30 grid, and (**b**) a 300 × 300 grid. *Blue* means coexistence of virus and cells. *Red* and *orange* indicate extinction of the cells and thus the virus. *Red* is used if extinction occurs before the boundary of the system has been reached, while *orange* is used if extinction occurs after cells have reached the boundary of the system. *Grey* indicates extinction of the virus while cells persist. Above the *white* line and below the *black* line, the "local" equilibrium number of uninfected and infected cells, respectively, is greater than one. Below the *yellow* line, the virus can successfully invade its target cell population. The capital letters indicate different spatial patterns that are described in the text and in Fig. 14.3. In these simulations, the probability for an uninfected cell to die was $D = 0$

Figures 14.3 and 14.4 summarize the observed spatial and temporal patterns. Note that identical results were observed over a very large range of initial conditions (differences were only observed if the initial number of cells was such that immediate stochastic extinction is likely, which are not regions of interest with respect to our study). Also, changing the stopping time of the simulations did not yield a significant change over several orders of magnitude of the time variable. In the next sections, we examine the outcomes first in a relatively small 30 × 30 grid, and subsequently in a larger, 300 × 300 grid.

14.4.1 Small Grid

In the 30×30 grid, the following outcomes are found (Fig. 14.2a, and Figs. 14.3 & 14.4).

Two types of target cell extinction can be observed, both associated with the initial "hollow ring" structure. According to pattern **A**, *virus-mediated target cell extinction*, both the target cell and the virus populations spread outward in space as a wave, but the virus wave overtakes the target cell wave, leading to extinction of both populations. Pattern **B**, *boundary-mediated extinction*, represents weaker viruses compared to case A. In pattern B, the virus wave catches up with the target cell wave, leaves no uninfected cells behind in its wake, but fails to eliminate the target cell wave. Instead, the two waves travel together with the same velocity until the boundary in reached. The target cell population can escape the virus only by spreading outward. Once the boundary is reached, this is not possible anymore, explaining the extinction. Note that although real tumors are capable of breaking out of homeostatic control and spreading beyond the "carrying capacity" of their environment, boundary-mediated extinction can still take place. Genetic transformations associated with waves of clonal expansion or induction angiogenesis generally happen on longer time scales. Therefore, it is realistic to assume the existence of some geometric constraints (even temporary). Pattern B represents the situation where extinction is a consequence of such spatial constraints.

Another type of outcome is the coexistence of infected and uninfected cells, which is shown in pattern **C**, *constant density coexistence*. As the virus population spreads out in space, it leaves behind uninfected cells with a high probability, leading to the disperse pattern of initial expansion and the absence of any clear traveling waves. Instead, the expanding virus population leaves behind a mix of both populations, which eventually is found across the whole space and is characterized by an equilibrium density that is determined by the parameters of the system, while the populations settle around a stochastic steady state.

Finally, there are two types of virus extinction patterns. Pattern **D**, *virus extinction despite invasion*, represents a virus extinction regime where the virus can initially invade the target cell population, but does not persist in the long term. The virus reduces the target cell population, and subsequently goes extinct. This leaves the uninfected cell population to grow unopposed. The stronger the virus (lower value of A/B), the less likely this is observed in this regime, because the uninfected cell population is more likely to be driven extinct before the virus population hits extinction. The second extinction pattern **E**, *lack of invasion*, is observed when the virus population cannot invade the target cell population and goes extinct (spatial and temporal pattern not shown).

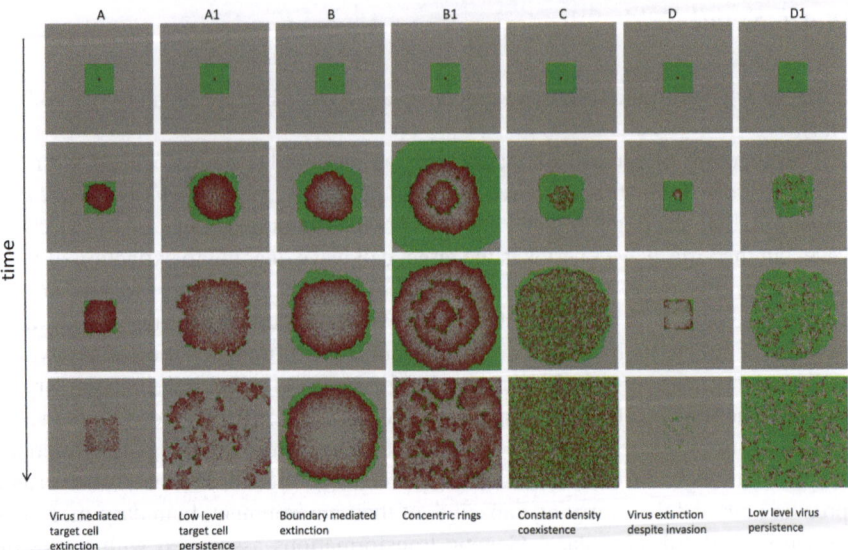

Fig. 14.3 Seven spatial patterns observed in the agent-based model. For each pattern, four snapshots in time are shown. *Green* indicates uninfected cells, *red* infected cells, and *grey* empty patches. See corresponding capital letters in Fig. 14.2, showing in which parameter regions the individual patterns are observed. The time series that are associated with the individual patterns are shown in Fig. 14.4. See text for details. The simulations were run on a 300 × 300 grid. The simulations were started by placing a small number of infected cells (5 × 5 cells) into a larger space filled with infected cells (13 × 13 cells). Parameters were chosen as follows: (**a**) $R = 0.013$; $D = 0$; $B = 0.14$; $A = 0.003$; **a1** $R = 0.15$; $D = 0$; $B = 0.32$; $A = 0.007$; (**b**) $R = 0.014$; $D = 0$; $B = 0.015$; $A = 0.00056$; **b1** $R = 0.04$; $D = 0$; $B = 0.032$; $A = 0.0016$; (**c**) $R = 0.014$; $D = 0$; $B = 0.032$; $A = 0.008$; (**d**) $R = 0.0002$; $D = 0$; $B = 0.019$; $A = 0.0032$; **d1** $R = 0.069$; $D = 0$; $B = 0.64$; $A = 0.18$

14.4.2 Analytical Insights: Global Outcome can be Predicted by the Local Dynamics

Although the spatial stochastic predator-prey system studied here exhibits a variety of patterns, its dynamics can be understood by studying the local interactions of the agents. In this section, we show that the global outcomes of the spatially distributed system can be predicted by utilizing the laws of local dynamics. We start from the well-known system of ordinary differential equations that can be derived for our agent-based model if no spatial restrictions were in place, and reproduction and infection events were driven by laws of mass-action [1]:

$$\dot{x} = Rx\left(1 - \frac{x + y}{K}\right) - \frac{Bxy}{K}, \tag{14.1}$$

$$\dot{y} = \frac{Bxy}{K} - Ay, \tag{14.2}$$

where the number of uninfected cells is denoted by x, and the number of infected cells by y. In these equations, K has the meaning of carrying capacity. This well-known modified Lotka-Volterra system [4, 5] is characterized by two equilibria: (i) The uninfected population persists at carrying capacity, while the virus population is extinct, i.e., $x^{(0)} = K$, $y^{(0)} = 0$; (ii) Alternatively, the virus establishes a successful infection, such that $x^{(1)} = AK/B$, $y^{(1)} = RK(B - A)/B(R + B)$. The latter equilibrium is stable if the basic reproductive ratio of the virus is greater than one, which is equivalent to the inequality $A < B$. The approach to the coexistence equilibrium can be either monotonic, or can involve damped oscillations.

While these properties of the virus-cell system are well-known, it is usually thought that the ordinary differential equations can only be applied to a well-mixed system, and fail to describe a spatially distributed system of cells. Contrary to this, Fig. 14.2 demonstrates that, if interpreted correctly, the above system can explain a lot of the patterns that arise in the spatial agent-based model. Let us think of the carrying capacity coefficient, K, as the size of the "local neighborhood" where cell-to-cell interactions happen in a spatial model. In our case, this neighborhood consists of $K = 9$ cells (a cell plus its eight nearest neighbors, the relevant characteristic scale of our spatial model). Model (14.1–14.2) with the modified parameter K is capable of informing us of the local equilibrium density of the infected and uninfected cells, which in term is correlated with the expected long-term behavior of the spatial system.

In Eqs. (14.1–14.2), the number of uninfected cells at the equilibrium (the value $x^{(1)}$) is proportional to K. In order for this equilibrium to be biologically meaningful, this value must be greater than one cell. The equation $x^{(1)} = 1$ defines the white line in Fig. 14.2. Similarly, the number of infected cells in local neighborhoods must be greater than one, which yields the black line, $y^{(1)} = 1$. We can see that the coexistence region in Fig. 14.2a (regime C) corresponds to the parameters for which both equilibrium values are larger than one; it is enclosed by the lines $x^{(1)} = 1$ and $t^{(1)} = 1$ obtained directly from the cancer-virus equations. The white line $x^{(1)} = 1$ outlines the lower boundary of the coexistence region, while the black line $y^{(1)} = 1$ defines the upper boundary. Note that a more precise definition of the upper bound of the coexistence region is given by the yellow line in Fig. 14.2a, below which the virus is strong enough to invade the cell population. This invasion threshold was determined by numerical simulations.

Thus, in the spatial system, target cell extinction is observed if the local equilibrium number of uninfected cells is less than one (regions A & B, Fig. 14.2a, below white line). Extinction of virus only is observed either following initial invasion if its local equilibrium is less than one (region D, Fig. 14.2a, area encased by white, black, and yellow lines) or if invasion is impossible (region E, Fig. 14.2a, above yellow line). The finding that equilibrium properties of simple ODE models that describe the dynamics in a small local neighborhood can predict the outcome of the spatial system has important practical implications. Note, however, that this method is unable to explain all the details of the diagram in Fig. 14.2. In particular, the proximity of the black ($I^{(1)} = 1$) line to the boundary between regions A and B is purely coincidental. The equilibrim analysis predicts extinction in regions A and B, but cannot distinguish between virus-mediated extinction (A) and boundary-mediated extinction.

Please note that the idea of a "characteristic scale" has been previously proposed in the literature in the context of different predator-prey models [6] where the system's behavior was found most predictable on an intermediate scale defined by the agents' motility and interactions. In [7], it was shown that in a class of systems exhibiting oscillatory dynamics, the functional forms governing the local predator-prey interactions at those characteristic scales are the same as the ones describing a perfectly mixed, mass-action system, but contain different parameters. This allowed the authors to approximate the long-term dynamics of the spatial system at large scales with a temporal predator-prey model describing local interactions. In the same spirit, the ideas presented is this section (and in [1]) allow us to learn about the long-term outcomes of a spatial system by studying its local behavior.

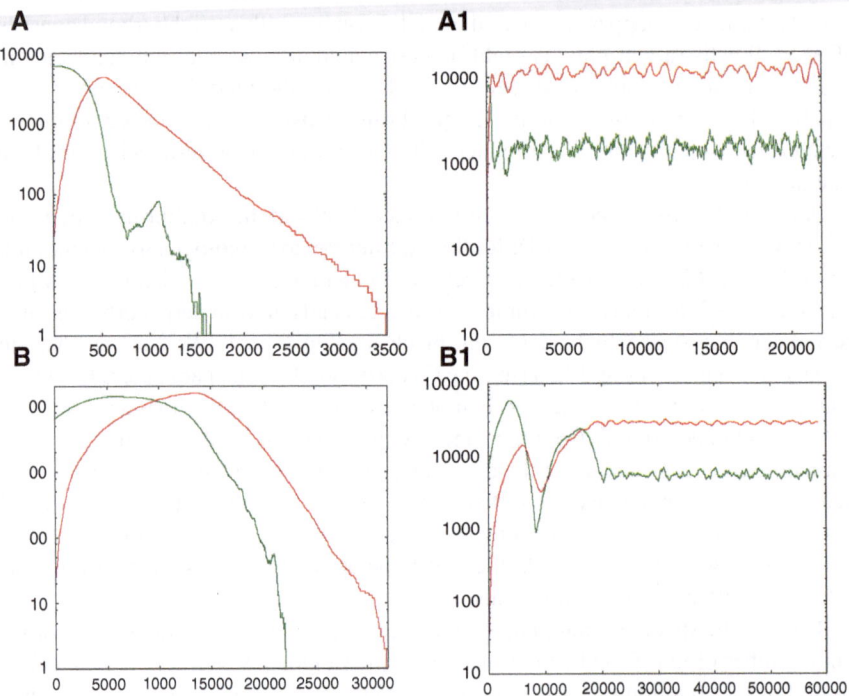

Fig. 14.4 Typical time series corresponding to the spatial patterns presented in Fig. 14.3, based on a single run of the spatial agent-based model, assuming a 300 × 300 grid. See text for details. Parameter values and initial conditions are given in Fig. 14.3

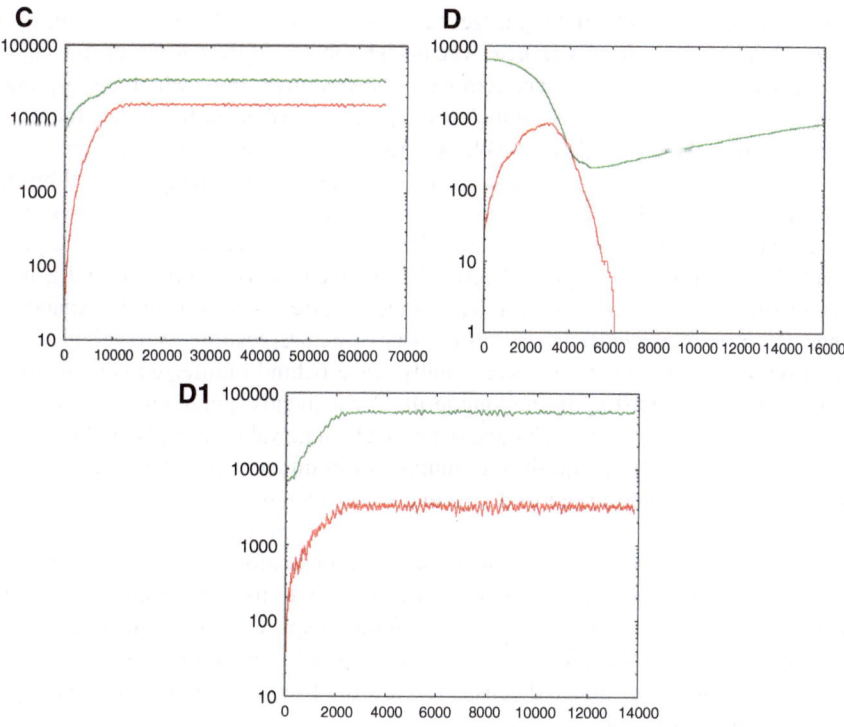

Fig. 14.4 (continued)

14.4.3 Large Grid

In a larger, 300×300, grid (Fig. 14.2b, and Figs. 14.3, 14.4), the basic patterns found in a small grid are still in place, but additional complexity is observed. In the parameter space where target cell extinction happens in the smaller grid, regions of coexistence can occur. In pattern **A1**, the expanding virus wave proceeds initially as a "hollow ring" structure, catches up with the target cell wave, leaves no uninfected cells in its wake, but only partially breaks the target cell wave. The virus is not efficient enough to eliminate the target cell wave, as observed in pattern A, but still strong enough not to leave it intact, as observed in pattern B. The partially broken wave structure allows the uninfected cells to escape not only outward, but in all directions. Hence, local extinction combined with continuous target cell movement away from the virus leads to persisting moving fronts, which can go extinct and give rise to new fronts over time. Thus, more extensive population fluctuations are observed in the long run (Fig. 14.4). This is the well-known regime of global persistence despite local extinction which is an important basis for the argument that space promotes coexistence [8]. The levels at which the uninfected cell population persists, however, are relatively low (Fig. 14.4). A sufficiently large grid size is required to observe this behavior, such that enough space is available for the moving target cell fronts to persist. We refer to pattern A1 as *low-level target cell persistence*. Region **B1** shows a different reason for target cell persistence at low levels, a pattern we call *concentric rings*, which corresponds to the concentric ring pattern of initial virus spread described earlier. When the virus wave expands, the probability to leave behind uninfected cells is proportional to the local equilibrium number of uninfected cells. In the region where this equilibrium number is just slightly below one, this does not occur often enough to be observed on a small grid. On a larger grid, however, it can be observed. These infrequent events lead to renewed target cell growth, followed by virus growth, and a new wave structure is formed. This can lead to the occurrence of concentric expanding rings. With time, stochasticity breaks the ring structure, leading to traveling fronts that eventually go extinct, but occasionally leave behind uninfected cells to form new fronts, thus persisting in the long term. Consequently, populations show more extensive fluctuations around characteristic steady-state values (Fig. 14.4). For lower values of A/B, the local equilibrium number of uninfected cells becomes too low for this to be observed in the grid size under consideration. Finally, in region **D1**, *low-level virus persistence*, global persistence of the virus despite local extinction is observed, leading to relatively strong population fluctuations (Fig. 14.4). While the virus invades the target cell population, it converges to its local equilibrium value that is less than one. However, movement through space before extinction occurs allows coexistence if the grid is sufficiently large. For lower values of R/B, the local equilibrium number of infected cells is too low to observe this outcome even in the context of the larger grid.

This analysis shows that increasing the grid size allows more complex outcomes to occur and increases the parameter region in which the cell populations persist. The additional patterns that emerge in larger grids are variations of those found in the smaller grid and involve nonequilibrium persistence, where extinction occurs locally, but movement through space allows cells to temporarily avoid extinction. These dynamics are well documented in the ecological literature [8]. Besides allowing cells to move through space, a larger grid size also increases the chances that certain rare events can occur. For example, boundary-mediated extinction (pattern B, Fig. 14.2) is less likely to occur in large grids. The larger the grid the higher the probability that uninfected cells are left in the core of the ring before the uninfected cell population has moved to the boundary and is eliminated by the virus. All these nonequilibrium persistence outcomes in larger grids, however, are characterized by persistence of the cells at very low levels, which can be considered controlled persistence and does not involve uncontrolled cellular growth. Therefore the outcome can still be predicted by the "local mass action equilibrium values" discussed above: if the local equilibrium of uninfected cells, predicted by the ODEs, is less than one, we can expect either extinction or controlled persistence. If the local equilibrium of uninfected cells is greater than one, we can expect to see uncontrolled cellular growth. The lower the local equilibrium of uninfected cells the less likely controlled persistence occurs and the more likely extinction is observed. However, this could not be demonstrated systematically for larger grids due to the extensive computational costs involved.

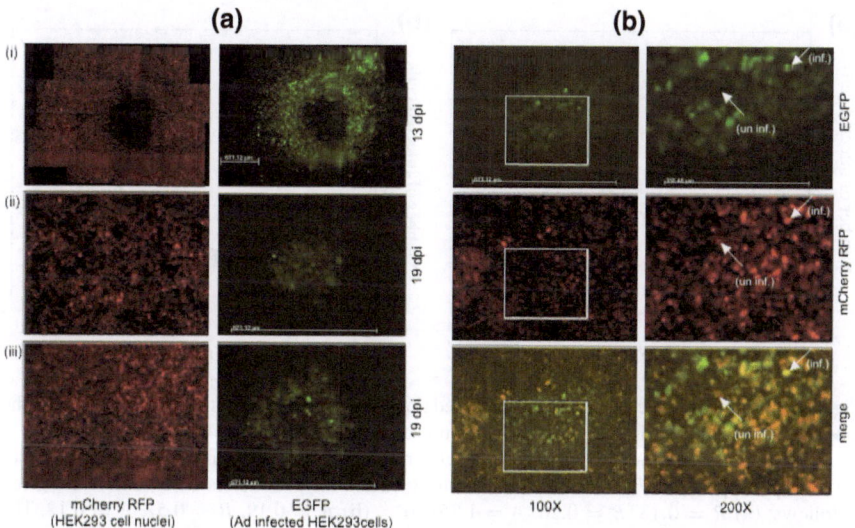

Fig. 14.5 Experimentally observed spatial patterns of virus infection, using the adenovirus AdEGF-Puci on HEK293-mCherry cells [1]. (**a**) Three patterns were observed: a hollow ring (*top*), a disperse pattern (*middle*), and a ring filled with uninfected cells that is eventually predicted to develop into a disperse pattern (*bottom*). Infected cells are shown in *green* on the *right*, and all cells (uninfected + infected) in *red* on the *left*. (**b**) Closer examination of the disperse, or limited, growth pattern, and magnified to different degrees (100x and 200x). The *top* panels show infected cells in *green*, the middle panels show all cells in *red*. Because the mCherry *red* fluorescent protein cannot distinguish between infected and uninfected cells, the *red* and the *green* images are merged in the *bottom* panel, illustrating infected versus uninfected cells. The arrows in the *right* panels point to an infected cell (inf.) and an uninfected cell (un inf.) within the center of the virus infected region of the cells

The long-term outcomes shown in Fig. 14.3 are obviously related to the initial growth patterns described in Fig. 14.1. Patterns A, A1, and B in Fig. 14.3 arise out of the hollow-ring structure. Pattern B1 in Fig. 14.3 emerges from the concentric ring structure. Patterns C, D, and D1 are consequences of the disperse growth pattern/filled ring structure, which for faster viral spread rates typically leads to coexistence of the virus and cell populations, while extinction of the virus population can be observed for smaller virus spread rates.

14.5 Experimentally Observed Patterns of Virus Spread

In order to experimentally examine spatial virus spread in a setting corresponding to the assumptions of our model, a recombinant adenovirus type-5 (Ad5) was constructed that expresses enhanced jellyfish green fluorescent protein (EGFP), AdEGF-Puci, and grows on human 293 embryonic kidney epithelial (293) cells [9]. The experiment was set up such that cells are arranged in a two-dimensional layer, and

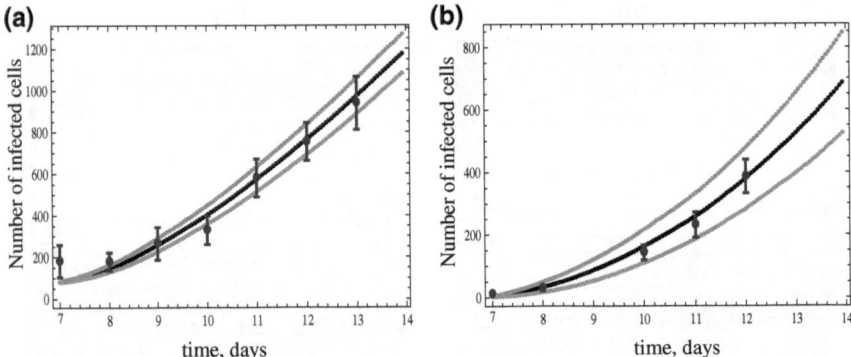

Fig. 14.6 Two different experimentally observed time series of adenovirus infection, fitted with the model. The *black middle* line through the data represents the time series predicted by the agent-based model. The *upper* and lower lines show the standard deviations. (**a**) A ring structure (Fig. 14.5a). (**b**) A disperse pattern (Fig. 14.5b). The best-fitting values for parameters R, B, and A are given as follows: (**a**) $R = 0.18$, $B = 0.26$, $A = 1.85x10^{-2}$. (**b**) $R = 0.19$, $B = 0.52$, $A = 0.12$. The parameter D was kept constant at $D = 0$

virus spread is most likely to occur to neighboring cells. An agar overlay prevents long-range spread of the virus away from infected cells in the culture medium. This setup allows to quantify not only the number of infected cells over time, but also the spatial patterns of infected cells that are formed as the virus population expands [1]. In addition, fluorescent markers were used to visualize the spatial distribution of all cells (infected and unifected) by generating HEK293-H2BmCherry cells, that stably express the core nuclear histone protein H2B fused to mCherry (a highly photostable, monomeric red fluorescent protein (RFP)) [10]. Thus, using HEK293-H2BmCherry cells allows visualization of all the cell nuclei (i.e., intact cells) in any particular culture. The culture was infected at a very low multiplicity of infection (MOI), such that any area of infection resulted from a single "founder" infected cell. Each culture contained several such founder cells that were sufficiently separated from each other, to allow tracking multiple growth foci across the dish. The earliest stages of virus growth starting from a single founder infected cell were described in [9]. In [1], we followed the growth of such spreading infections and characterized the consequent growth patterns. We observed three basic patterns of virus spread, which correspond to the patterns predicted by the model. The experimentally observed patterns are shown in Fig. 14.5a and described as follows. (i) In the first pattern, the virus infection spreads rapidly outward as a ring, leaving no cells behind in the core of the ring (Fig. 14.5a, pattern (i)). This classic plaque pattern is observed in virus growth experiments, and it corresponds to the hollow ring structure predicted by mathematical modeling above. In the second and third patters there is viral spread, but it is limited. (ii) In the second case, a disperse growth pattern is observed, where the virus population expands as a mixed cluster of infected and uninfected cells (Fig. 14.5a, pattern (ii)). Finally, the virus population expands as a thinner ring, but in contrast

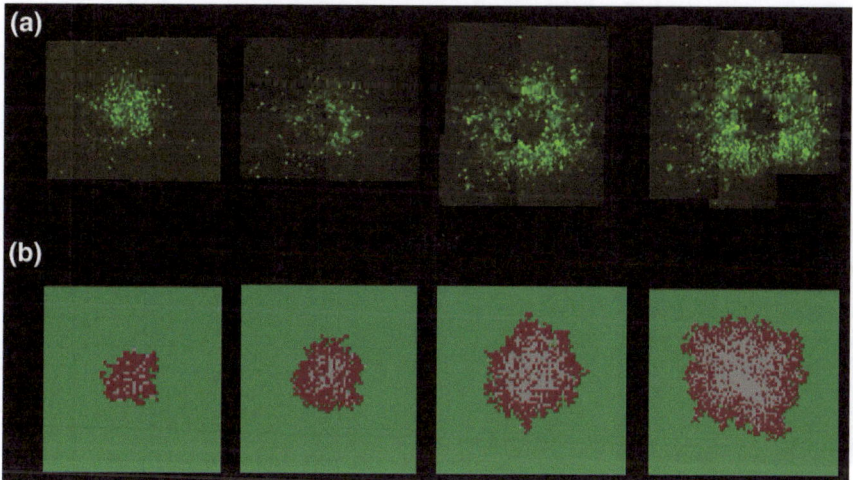

Fig. 14.7 Observed (**a**) and predicted (**b**) spatial pattern of adenovirus (AdEGFPuci) growth for the experiment that exhibits a ring structure (time series given in Fig. 14.3a). The predicted spatial pattern is the result of an individual run of the agent-based model with the parameter combination obtained from the model fitting procedure. Snapshots in time are shown, representing days 7, 9, 11, and 13 post infection. (**a**) The area of green fluorescence is shown, expressed by the infected cells, thus documenting the spatial spread of the virus through the population target cells arranged in a two-dimensional setting. (**b**) In the computer simulation, *green* indicates infected cells, *red* infected cells, and *grey* empty spots

to the first case, uninfected cells are left behind in the core of the ring (Fig. 14.5a, pattern (iii)). A limited growth pattern is magnified in Fig. 14.5b, in which uninfected cells are visible within the center of the virus infected population. In the top right panel of Fig. 14.5b, an AdEGFPuci infected (fluorescent) cell is indicated (arrow marked with "inf."), whereas an uninfected cell in the center of the spreading infection does not fluoresce green (arrow, un inf.). The same cells are indicated in the middle right panel of Fig. 14.5b, showing red fluorescence. In the bottom left panel of Fig. 14.5b, images of the top and middle panels are merged; infected cell (arrow, "inf.") fluorescence yellow, while the uninfected cell, (arrow, "un inf.") remains red. As mentioned, the area over which the infection spread, remained limited in patterns (ii) and (iii) and persisted throughout the infection (through 19 dpi). In contrast, in pattern (i), the ring of infected cells continued to spread outward as long as there was space; cell clearing in the center of the plaque was apparent at 13 dpi, as shown in Fig. 14.5a. Similar patterns of spreading infection were also seen in Ad293 cells, a HEK293 cell derivative optimized for adenovirus plaque assays. Overall, among 436 scored growth foci, the hollow ring structure was found in 45 %, and the limited patterns in 55 % of cases.

In order to go beyond the qualitative comparison of model and data, we fit the model to two sets of experimental data, one showing an expanding hollow ring, and the other the disperse growth pattern, see Fig. 14.6. The number of cells was

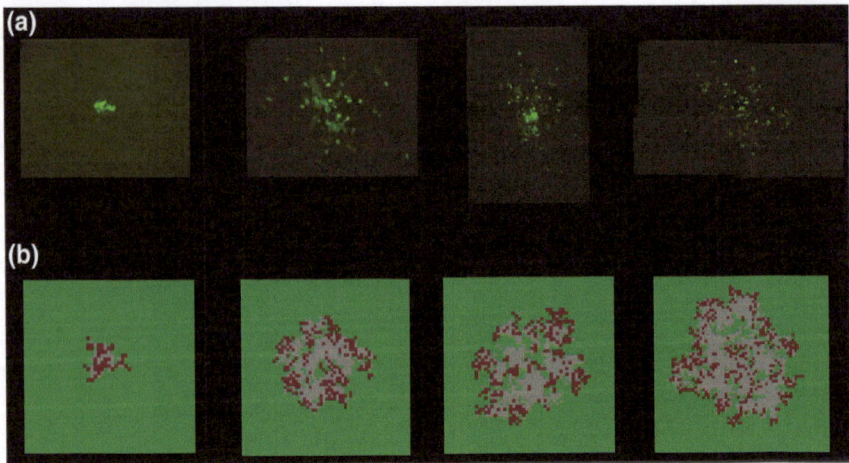

Fig. 14.8 Observed (**a**) and predicted (**b**) spatial pattern of adenovirus (AdEGFPuci) growth for the experiment that exhibits a disperse growth pattern (time series given in fig. onco4spsfig3b). The predicted spatial pattern is the result of an individual run of the agent-based model with the parameter combination obtained from the model fitting procedure. Snapshots in time are shown, representing days 7, 10, 11, and 12 post infection. (**a**) The area of *green* fluorescence is shown, expressed by the infected cells, thus documenting the spatial spread of the virus through the population target cells arranged in a two-dimensional setting. (**b**) In the computer simulation, *green* indicates infected cells, *red* infected cells, and *grey* empty spots

experimentally determined by measuring the fluorescent area of the infected cell population, divided by the fluorescent area of individual infected cells, using Photoshop. Each graph in Fig. 14.6 is based on a single experimental run. The area, however, was measured independently four times, giving rise to the plotted error bars. The black middle line through the data represents the time series predicted by the agent-based model, using a parameter combination that was obtained by a least squares fitting procedure. Since the model is stochastic, the predicted time series represents the average over 1,000 instances of the simulation. The upper and lower lines show the standard deviations added to and subtracted from the average. The experiment fitted in 14.6a shows virus growth characterized by the formation of a ring structure. Consequently there is a relatively short phase of quadratic growth, followed by a transition to linear growth. The experiment fitted in 14.6b shows disperse virus growth characterized by quadratic growth throughout time. The corresponding observed and predicted spatial patterns are shown in Figs. 14.7 and 14.8. The types of spatial patterns that emerged matched the observed ones qualitatively (Figs. 14.7 and 14.8). Please note that although this procedure found best fitting parameter values, their biological meaning remains questionable, since different parameter combinations can give rise to similarly good fits.

While the experimentally observed spatial patterns correspond to predicted ones, the experiments give rise to further observations that are not seen in the model and that are likely due to additional biological process that are at work in this in vitro system and that are not part of the model. The most puzzling observation was that identical experimental conditions, using the same virus-target cell system, gave rise to different patterns of virus growth. This indicates the existence of so far unidentified factors that influence virus spread in this in vitro system. It is possible that initial events, stochastic in nature, might determine the remaining fate of the virus population. One hypothesis is that infection of cells triggers the production of anti-viral factors such as interferon, by the infected cell, which could induce an anti-viral state in neighboring cells.

14.6 Summary

This chapter discussed a computational model that describes the spatial spread of viruses through a culture of target cells, where target cells are arranged in a two-dimensional monolayer, and viruses are likely to only pass the offspring on to directly neighboring cells. While many more complex interactions certainly apply to the *in vivo* spread of oncolytic viruses through spatially structured tumors, understanding this simple scenario forms the basis for any further, more complex explorations. As the results have shown, although the model outcomes fit experimental data well qualitatively and quantitatively, even such a simple experimental setup gives rise to dynamics that are rather complicated and that likely involve biological processes that go beyond the basic ones assumed in the model. A full understanding of those processes will be crucial before research can move on to more complex, *in vivo* systems.

Problems

14.1 Investigate the equilibria and their stability for system (14.1–14.2) (cf problem 10.1).

14.2 Research Project

Consider a stochastic process whereby cells on a grid of size K undergo a stochastic birth-death type process, see [1]. Grid points can be empty or occupied by one of two types of cells: infected and uninfected. Denote by i the number of uninfected cells, and by j the number of infected cells. With rate A, infected cells die. With rate R, uninfected cells attempt reproduction. To proceed with reproduction, a random spot on the grid is chosen. If the spot is empty, the cells' offspring is placed in that spot; otherwise, reproduction is aborted. Finally, with rate B, infected cells attempt to infect a susceptible cell. To proceed with infection, a random spot on the grid is chosen. If it contains an uninfected cell, the cell becomes infected; otherwise infection event does not happen.

(a) Formulate a stochastic process in the space (i, j) of infected and uninfected cells,

$i, j \geq 0, i + j \leq K$. Formulate the Kolmogorov forward equation for the function $\varphi_{i,j}(t)$, the probability to be in state (i, j).

(b) Derive ordinary differential equations for the average values, $\langle i \rangle = \sum_{i,j} \varphi_{i,j} i$ and $\langle j \rangle = \sum_{i,j} \varphi_{i,j} j$. How do these equations compare with equations (14.1–14.2)?

14.3 Research Project

Consider the following empirical model of infection spread. Assume that infection is introduced at the point with coordinate $(0, 0)$ at time $t = 0$. Suppose that the spread happens according to the law

$$\dot{\mathcal{A}} = \gamma \sqrt{\mathcal{A}},$$

where $\mathcal{A}$ is the area of infected plaque, γ is an empirical parameter, and the square root reflects the fact that the infection happens proportional to the circumference of the infected circle.

(a) Solve the above equation and find the radius of the plaque as a function of time. Resolve this for t as a function of r, to find out the time at which infection will spread to distance r from the center.

(b) Assuming that infected cells die at rate a, find the density of infection at time t as a function of the distance from the center, see [1].

(c) By integrating over the circle, find the total number of infected cells in a plaque as a function of time,

$$y(t) = \frac{\gamma^2}{4\pi} \frac{e^{-at} - 1 + at}{a^2}. \tag{14.3}$$

(d) Perform a similar calculation in 3D, to obtain

$$y(t) \propto \frac{2(1 - at - e^{-at}) + (at)^2}{a^3}. \tag{14.4}$$

14.4 (a) Using Eq. (14.3), it shows that the growth law for the number of infected cells in 2D is quadratic in time for small values of at, and the number if infected cells grows linearly with time for large values of at.

(b) What is the behavior for short and long times (small at and large at) in the case of 3D growth, Eq. (14.4)

References

1. Wodarz, D., Hofacre, A., Lau, J.W., Sun, Z., Fan, H., Komarova, N.L.: Complex spatial dynamics of oncolytic viruses in vitro: mathematical and experimental approaches. PLoS Comput. Biol. **8**(6), 547 (2012) (e1002)
2. Sato, K., Matsuda, H., Sasaki, A.: Pathogen invasion and host extinction in lattice structured populations. J. Math. Biol. **32**(3), 251–268 (1994)
3. Goodrum, F., Ornelles, D.: Roles for the e4 orf6, orf3, and e1b 55-kilodalton proteins in cell cycle-independent adenovirus replication. J. Virol. **73**(9), 7474–7488 (1999)

4. Nowak, M.A., May, R.M.: Virus dynamics: mathematical principles of immunology and virology Oxford University Press, Oxford (2000)
5. Anderson, R.M., May, R.M.: Infectious Diseases of Humans. Oxford University Press, Oxford (1991)
6. Deroos, A.M., Mccauley, E., Wilson, W.G.: Mobility versus density-limited predator prey dynamics on different spatial scales. Proc. R. Soc. Lond. B Biol. Sci. **246**(1316), 117–122 (1991)
7. Pascual, M., Mazzega, P., Levin, S.A.: Oscillatory dynamics and spatial scale: the role of noise and unresolved pattern. Ecology **82**(8), 2357–2369 (2001)
8. Hassell, M.: The Spatial and Temporal Dynamics of Host-Parasitoid Interactions. Oxford University Press, Oxford (2000)
9. Hofacre, A., Wodarz, D., Komarova, N.L., Fan, H.: Early infection and spread of a conditionally replicating adenovirus under conditions of plaque formation. Virology **423**(1), 89–96 (2012)
10. Shaner, N.C., Campbell, R.E., Steinbach, P.A., Giepmans, B.N.G., Palmer, A.E., Tsien, R.Y.: Improved monomeric red, orange and yellow fluorescent proteins derived from discosoma sp red fluorescent protein. Nature Biotechnology **22**(12), 1567–1572 (2004)

Chapter 15
Oncolytic Viruses and the Eradication of Drug-Resistant Tumor Cells

Abstract The main objective of oncolytic virus therapy described so far is the successful control or elimination of the tumor. Here, we describe an alternative goal of virus therapy in case the virus fails to eradicate the tumor. The virus can be used to "pre-treat" the cancer and specifically drive drug-resistant mutants extinct. This prepares the ground for subsequent drug treatment, which can then successfully drive the tumor into remission without resistance-induced failure. The key to this concept is that many drug-resistant mutants suffer a fitness cost compared to the susceptible cells in the absence of the drug. When the virus is introduced before the drug, then the less fit resistant cells share an enemy, the virus, with the fitter susceptible cells. In this case, apparent competition can occur, which can lead to exclusion of the inferior type even though coexistence occurs without the virus. Therefore, apparent competition, mediated by the oncolytic virus, can drive drug-resistant mutant cells extinct that would otherwise be present before treatment, and prepare the tumor for subsequent successful drug therapy.

Keywords Oncolytic viruses · Apparent competition · Drug resistant mutants · Fitness cost · Selective advantage · Susceptible population · Pre-treatment · Wild-type cells · Ordinary differential equations · Coexistence · Competitive exclusion · Extinction · Drug therapy · Combination therapy

15.1 Introduction

All chapters on oncolytic viruses so far have examined the conditions under which the virus can drive the cancer cell population extinct under various biological assumptions. Indeed, this is the ultimate goal of virus therapy. However, if the virus falls short of eradicating or controlling the cancer cell population, leading to continued growth, virus therapy might still be useful at making the cancer more susceptible to subsequent chemotherapy or to treatment with small molecule inhibitors. This is the

N. L. Komarova and D. Wodarz, *Targeted Cancer Treatment in Silico*, 215
Modeling and Simulation in Science, Engineering and Technology,
DOI: 10.1007/978-1-4614-8301-4_15, © Springer Science+Business Media New York 2014

focus of the current chapter, and this also bridges with the first part of the book, which considers cancer cell dynamics during treatment with small molecule inhibitors. A prevalent problem both with chemotherapy and with small molecule inhibitors is the presence of drug-resistant cells. Theoretical and empirical work strongly suggests that resistant mutants pre-exist at low levels before the start of treatment and are not generated during therapy itself [1–6]. Upon treatment, the resistant cells obviously enjoy a selective advantage and rise at the expense of the susceptible cells. The virus population typically infects both drug-resistant and susceptible cell populations equally well. Therefore, these two cell populations share a natural enemy, the virus. In this setting, the phenomenon of "apparent competition" can apply [7]. This means that even if the drug-susceptible and resistant cell populations are not in direct competition with each other and/or coexist, the sharing of a natural enemy can lead to outcomes that are identical to competitive exclusion, i.e., the fitter cell population excludes the less fit one. The fitter cell population can produce an amount of virus that is too much for the less fit cell population to survive. If the resistant cell population is less fit than the susceptible one, virus therapy can lead to the extinction of otherwise pre-existing resistant cell populations. Therefore, another use of oncolytic viruses could be to "pre-treat" the cancer with the virus, driving drug-resistant cell populations extinct, and then treating the cancer with small molecule inhibitors or chemotherapy. This is explored in this chapter with mathematical models.

15.2 The Model

We consider a mathematical model that describes the spread of an oncolytic virus through a population of tumor cells and that assumes the existence of two subpopulations of cancer cells: cells that are susceptible to a given drug (from now on called wild-type cells), and cells that are resistant against this drug. The model includes the following variables: uninfected wild-type cells, x_1; uninfected drug-resistant cells, x_2; infected wild-type cells, y_1; infected drug-resistant cells, y_2. The model does not take into account a free virus population explicitly. Because the turnover of virus populations is much faster than that of infected cells, we assume that the free virus population is in a quasi-steady state.

Uninfected wild-type cells grow with a rate

$$\frac{r_1 x_1 (1 + \eta)}{x_1 + x_2 + y_1 + y_2 + \eta}.$$

Thus, we assume saturated growth. The lower the total number of cells, the more this term approaches exponential growth. The higher the number of cells, the more this term approaches linear growth. The parameter η determines at what tumor size the exponential growth starts to saturate. For high values of η, the behavior converges to an exponential growth model. For low values of η, the behavior converges to a linear growth model. We assume that growth does not stop at a given tumor size, as would be the case with logistic growth. The reason is that we aim to model the most aggressive tumor growth, where a relatively fast expansion of the tumor cell population is observed until the organism dies. Saturated growth is assumed because several constraints, such as space, are likely to prevent straight exponential growth at large numbers of cells.

Upon contact with virus, wild-type cells become infected with a rate β_1. Traditional virus dynamics models assume straightforward mass-action [8, 9], where the rate of infection is directly proportional to the number of uninfected and infected cells. However, this leads to "boom and bust" dynamics that do not reflect the acute dynamics of the infection in a realistic way. In particular, the initial spread of the virus depends heavily on the number of target cells, which can be huge in the context of a tumor. Alternatively, frequency-dependent transmission can be assumed [10–14]. In this case, the rate of infection is not proportional to the number of infected cells, but to the fraction of infected cells in the tumor population. This gives rise to the opposite result that the initial virus spread is independent of the number of target cells. An intermediate scenario [15–17] is most likely to be realistic, and this is what we assume in our model. The transmission term is thus given by

$$\frac{\beta_1 x_1 (y_1 + y_2)(1 + \varepsilon)}{x_1 + x_2 + y_1 + y_2 + \varepsilon}.$$

The parameter ε determines how much the model approaches the mass-action or frequency dependence assumption. The higher the value of ε, the less the transmission is frequency dependent.

The drug-resistant cancer cell population is modeled according to the same principles. It grows with a rate $r_2 x_2 (1 + \eta)/(x_1 + x_2 + y_1 + y_2 + \eta)$. Drug-resistant cells become infected with a rate $\beta_2 x_2 (y_1 + y_2)(1 + \varepsilon)/(x_1 + x_2 + y_1 + y_2 + \varepsilon)$. In addition, we assume that drug-resistant cells are produced by mutation from wild-type cells with a rate $u x_1$.

Infected cells are assumed not to divide, and die with a rate a_1 and a_2, respectively. The model is given by the following set of ordinary differential equations:

$$\dot{x}_1 = \frac{r_1 x_1 (1 + \eta)}{x_1 + x_2 + y_1 + y_2 + \eta} - \frac{\beta_1 x_1 (y_1 + y_2)(1 + \varepsilon)}{x_1 + x_2 + y_1 + y_2 + \varepsilon} - u x_1, \tag{15.1}$$

$$\dot{x}_2 = \frac{r_2 x_2 (1 + \eta)}{x_1 + x_2 + y_1 + y_2 + \eta} - \frac{\beta_2 x_2 (y_1 + y_2)(1 + \varepsilon)}{x_1 + x_2 + y_1 + y_2 + \varepsilon} + u x_1, \tag{15.2}$$

$$\dot{y}_1 = \frac{\beta_1 x_1 (y_1 + y_2)(1 + \varepsilon)}{x_1 + x_2 + y_1 + y_2 + \varepsilon} - a_1 y_1, \tag{15.3}$$

$$\dot{y}_2 = \frac{\beta_2 x_2 (y_1 + y_2)(1 + \varepsilon)}{x_1 + x_2 + y_1 + y_2 + \varepsilon} - a_2 y_2. \tag{15.4}$$

The mathematical properties of the current model are explored as follows. First, we assume that the mutation rate $u = 0$. We simply assume initial conditions in which both cell populations are present. This assumption will be subsequently relaxed.

Fig. 15.1 The growth of wild-type and drug-resistant mutants (**a**) in the absence of an oncolytic virus, and (**b**) in the presence of an oncolytic virus, assuming absence of mutations. In the absence of the virus, both cancer cell populations grow without bound. In the presence of the virus the wild-type cancer cells approach an equilibrium because the virus prevents unlimited growth of the cancer cells. The virus, however, fails to eradicate the cancer. The drug-resistant cancer cells go extinct because of apparent competition mediated by the oncolytic virus. Parameters were chosen as follows: $r_1 = 7$, $r_2 = 5$, $\beta_1 = \beta_2 = 0.1$, $a_1 = a_2 = 0.5$, $\varepsilon = 10$, $\eta = 10^9$, $u = 0$

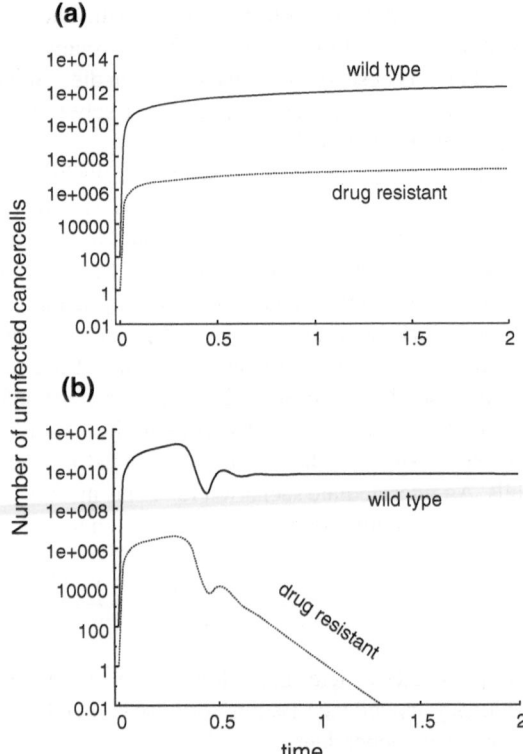

15.3 Dynamics in the Absence of Mutations

In the absence of any virus, both types of cancer cells can grow to ever-increasing levels (Fig. 15.1a). This is because no upper bound to population growth is assumed. Consequently, both the resistant and susceptible cells coexist during growth, even though the model assumes that the two cell populations influence each others' growth rate to a certain degree. In the following the assumption will be made that the virus can grow successfully in both the wild-type and the resistant cell populations alone. That is,

$$(1 + \varepsilon)\beta_1/a_1 > 1 \text{ and } (1 + \varepsilon)\beta_2/a_2 > 1,$$

given that the number of cells is large. Assuming that the virus fails to eliminate the cancer, the following behavior is observed in the presence of the virus. The virus infection prevents unbounded cancer growth and the overall number of cancer cells converges toward an equilibrium (Fig. 15.1b). The virus population also converges to an equilibrium, which is not shown in the graph. However, only one tumor cell variant converges to the equilibrium, and the other goes extinct (Fig. 15.1a). Coexistence is not possible in the presence of the virus. The wild-type cells persist and the resistant cells go extinct if

$$\frac{r_1}{\beta_1} > \frac{r_2}{\beta_2}. \tag{15.5}$$

The resistant cells persist and the wild-type cells go extinct in the opposite case, if inequality (15.5) is reversed. The equilibrium expressions for these two outcomes are given by lengthy second degree polynomial expressions which are not written here for simplicity. Data indicate that drug-resistant mutations can confer a fitness cost to the cells in the absence of therapy in a variety of contexts. As discussed in Chap. 4 in the context of CML, some drug-resistant mutations can grow less efficiently than sensitive cells [18]. If we assume that the resistant cancer cells carry a fitness cost compared to the wild-type cells (i.e., they grow slower, $r_2 < r_1$, and if we assume that the replication rate of the virus is identical in wild-type and resistant cells ($\beta_1 = \beta_2$), then the only possible outcome of the virus infection is that the resistant cancer cell population is driven extinct (Fig. 15.1b). Although the two cancer cell populations do not directly compete for a shared resource in the model, the presence of the virus leads to competitive exclusion, where the fitter type displaces the less fit type. The fact that "competitive exclusion" can occur if two species (that do not compete directly) share a pathogen is referred to as "apparent competition" in the ecological literature [7]. Apparent competition has been shown to be a potentially important factor that shapes ecological assemblages [19–22].

In the above analysis, it was assumed that the virus can potentially establish an infection in either of the cell types (wild-type or resistant) alone. This is the most realistic parameter regime. If the virus replicates too slowly to establish an infection in either the wild-type or the resistant cell population alone, then it will not be possible to eradicate the population of the resistant cancer cells through virus-mediated competition. Therefore, the dynamics of this scenario is not further considered.

15.4 Effect of Mutations

The above analysis assumed that wild-type cells do not give rise to resistant cells by mutation ($u = 0$) and that a certain number of resistant cancer cells were present at the beginning of the computer simulations. Here, this assumption is relaxed. We can set the mutation rate $u = 10^{-7} - 10^{-9}$ according to the physiological mutation rate, although it can be higher if the cancer cells have some form of genetic instability. In particular, in the context of chromosomal instability, the mutation rate can be around $u = 10^{-4}$ if resistance is caused by chromosome duplications [23]. Now, the resistant cancer cells will not be driven extinct anymore. The dynamics are shown in Fig. 15.2. As before, the virus infection prevents further growth of the tumor cell population but fails to drive the tumor to low numbers. The susceptible cell population converges to an equilibrium. Because the resistant mutant has a lower fitness, it declines. However, the population levels do not go to zero because mutations received by the susceptible cells continuously give rise to resistant cells. Hence, the resistant cell population is

Fig. 15.2 Dynamics in the presence of the virus, assuming that susceptible cells mutate to become resistant. (a) A less effective and (b) a more effective virus are considered. The virus prevents continuous cell growth and the system converges to an equilibrium. Resistant mutants are maintained by a balance between their selective disadvantage and their generation by mutation from susceptible cells. The lower their level, the higher the chances that they will not be present in a stochastic setting. Parameters were chosen as follows: $r_1 = 7$, $r_2 = 5$, $a_1 = a_2 = 0.5$, $\varepsilon = 10$, $\eta = 10^9$, $u = 10^{-9}$. For (a) $\beta_1 = \beta_2 = 0.1$. For (b) $\beta_1 = \beta_2 = 0.3$

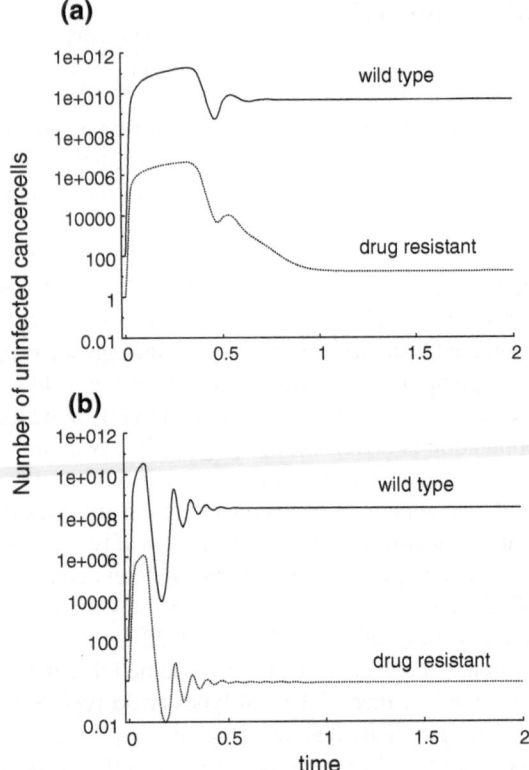

maintained at a level determined by the balance between the selective disadvantage and the influx by mutation (selection-mutation balance). The exact level at which the resistant mutant is expected to persist is determined by model parameters (Fig. 15.2). If the level is sufficiently low, this could correspond to extinction in practical terms due to stochastic effects. Lower levels of the resistant mutant are promoted by lower mutation rates, the effectiveness of the oncolytic virus and thus the equilibrium level of susceptible cells, and by a higher fitness cost of the drug-resistant mutant. In order to determine whether the virus can eliminate the resistant cell population, these metrics would need to be known, and the outcome would need to be studied in the context of a stochastic version of this model. What is certain, however, is that instead of a continuously growing resistant cell population, the abundance of these cells remains stable at significantly reduced levels, thus greatly lowering the probability that they will be present when drug therapy is started. This underlines the notion that the virus can be used to "pre-treat" the cancer and to make it more susceptible to subsequent drug therapy. This is illustrated by computer simulations in Fig. 15.3. In Fig. 15.3a, parameters are such that the virus does not reduce the drug-resistant cell population below one before drug treatment is started. If the resistant

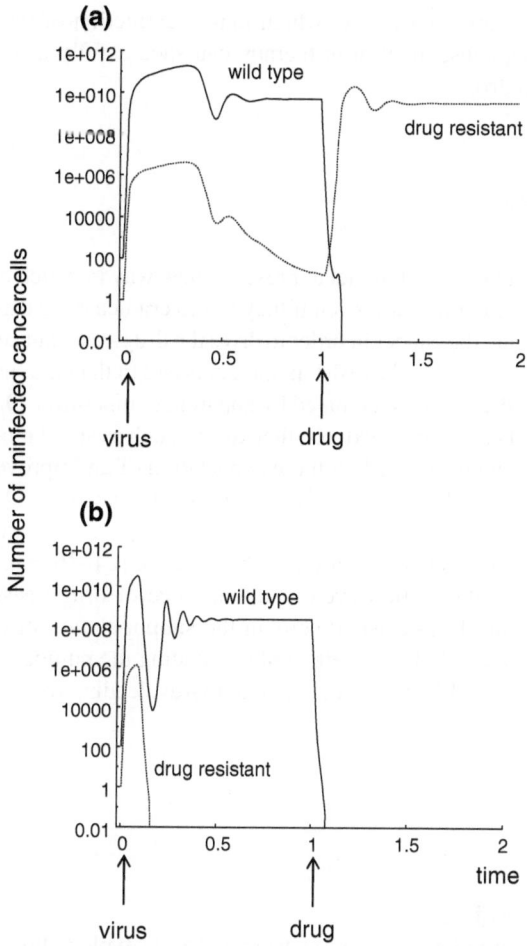

Fig. 15.3 Simulation of the sequential administration of oncolytic virus therapy and drug therapy, assuming that wild-type cancer cells can produce drug-resistant cells by mutation. Administration of virus and drug are indicated by *arrows*. (**a**) If resistant mutant levels are still relatively high, they are likely to exist at the start of drug therapy, leading to treatment failure. (**b**) If apparent competition lowers the resistant population more, they are more likely to be extinct upon start of drug therapy, which is then successful. Parameters were chosen as follows: $r_1 = 7$, $r_2 = 5$, $a_1 = a_2 = 0.5$, $\varepsilon = 10$, $\eta = 10^9$, $u = 10^{-9}$. For (**a**) $\beta_1 = \beta_2 = 0.1$. For (**b**) $\beta_1 = \beta_2 = 0.3$. Drug therapy was modeled by assuming an additional death rate for susceptible cells, $d = 5$. Because ODEs are used, stochastic extinction (although extremely likely in scenario (**b**)) cannot be captured. Instead, we assumed that once the resistant mutant numbers dropped below a very low threshold level, we can set $x_2 = y_2 = 0$

cells are present at the start of drug therapy, drug treatment will fail. In contrast, Fig. 15.3b shows a parameter region in which the virus reduces the drug-resistantcell population significantly below one, which makes extinction of resistant cells very likely. In this case, subsequent drug therapy can successfully drive the tumor into remission (Fig. 15.3b).

15.5 Summary

We have used a mathematical model to present a new way in which oncolytic viruses can be useful for cancer therapy, even if they fail to eradicate the cancer. A virus can be used to "pre-treat" the cancer in order to drive the drug-resistant cell clones toward extinction. This requires that the resistant mutant is less fit than the susceptible cells in the absence of the drug, and is mediated by apparent competition. Once the virus has driven the resistant cancer cells extinct, then drugs, such as small molecule inhibitors or chemotherapy, can be applied with the expectation of an improved response.

Note, however, that this argument does not take into account other complicating factors that can pose obstacles to successful drug therapy. For example, primitive or quiescent cancer cells that are not affected by the drug [24–26] can pose significant obstacles to cancer elimination even in the absence of drug-resistant mutants. In addition, if there are drug-resistant cells in the subpopulation of quiescent cancer cells, then the virus would be less efficient at reducing the number of resistant cells. Quiescent cells are not likely to produce much virus, leading to their survival rather than death.

Problems

15.1 Numerical project
Using numerical simulations of model (15.1–15.4), explore how the equilibrium number of drug-resistant cells depends on the infection rate of the virus, assuming that the infection rates are identical for wild-type and resistant cells ($\beta_1 = \beta_2$). Start with an infection rate such that the basic reproductive ratio of the virus is close to one (although greater than one), and then increase the basic reproductive ratio by raising the infection rate.

15.2 Research project
Find out more details about the concept of apparent competition. Describe how it is important in ecological interactions, and explore parallels between the ecological and the biomedical worlds.

References

1. Komarova, N.L., Wodarz, D.: Drug resistance in cancer: principles of emergence and prevention. Proc. Natl. Acad. Sci. U S A **102**(27), 9714–9719 (2005)
2. Diaz, L.A., J., Williams, R.T., Wu, J., Kinde, I., Hecht, J.R., Berlin, J., Allen, B., Bozic, I., Reiter, J.G., Nowak, M.A., Kinzler, K.W., Oliner, K.S., Vogelstein, B.: The molecular evolution of acquired resistance to targeted EGFR blockade in colorectal cancers. Nature **486**(7404), 537–540 (2012)
3. Montagut, C., Dalmases, A., Bellosillo, B., Crespo, M., Pairet, S., Iglesias, M., Salido, M., Gallen, M., Marsters, S., Tsai, S.P., Minoche, A., Seshagiri, S., Serrano, S., Himmelbauer, H., Bellmunt, J., Rovira, A., Settleman, J., Bosch, F., Albanell, J.: Identification of a mutation in the extracellular domain of the epidermal growth factor receptor conferring cetuximab resistance in colorectal cancer. Nat. Med. **18**(2), 221–223 (2012)
4. Maheswaran, S., Sequist, L.V., Nagrath, S., Ulkus, L., Brannigan, B., Collura, C.V., Inserra, E., Diederichs, S., Iafrate, A.J., Bell, D.W., Digumarthy, S., Muzikansky, A., Irimia, D., Settleman, J., Tompkins, R.G., Lynch, T.J., Toner, M., Haber, D.A.: Detection of mutations in EGFR in circulating lung-cancer cells. N. Engl. J. Med. **359**(4), 366–377 (2008)
5. Turke, A.B., Zejnullahu, K., Wu, Y.L., Song, Y., Dias-Santagata, D., Lifshits, E., Toschi, L., Rogers, A., Mok, T., Sequist, L., Lindeman, N.I., Murphy, C., Akhavanfard, S., Yeap, B.Y., Xiao, Y., Capelletti, M., Iafrate, A.J., Lee, C., Christensen, J.G., Engelman, J.A., Janne, P.A.: Preexistence and clonal selection of met amplification in EGFR mutant NSCLC. Cancer Cell **17**(1), 77–88 (2010)
6. Durrett, R., Moseley, S.: Evolution of resistance and progression to disease during clonal expansion of cancer. Theor. Popul. Biol. **77**(1), 42–48 (2010)
7. Holt, R.D.: Predation, apparent competition and the structure of prey communities. Theor. Pop. Biol. **12**, 197–229 (1977)
8. Wodarz, D.: Viruses as antitumor weapons: defining conditions for tumor remission. Cancer Res. **61**(8), 3501–3507 (2001)
9. Wodarz, D.: Gene therapy for killing p53-negative cancer cells: use of replicating versus non-replicating agents. Hum. Gene Ther. **14**(2), 153–159 (2003)
10. Antonovics, J., Iwasa, Y., Hassell, M.P.: A generalized model of parasitoid, venereal, and vector-based transmission processes. Am. Nat. **145**, 661–675 (1995)
11. Begon, M., Hazel, S.M., Baxby, D., Bown, K., Cavanagh, R., Chantrey, J., Jones, T., Bennett, M.: Transmission dynamics of a zoonotic pathogen within and between wildlife host species. Proc. Biol. Sci. **266**(1432), 1939–1945 (1999)
12. May, R.M., Anderson, R.M.: Transmission dynamics of HIV infection. Nature **326**(6109), 137–142 (1987)
13. McCallum, H., Barlow, N., Hone, J.: How should pathogen transmission be modelled? Trends Ecol. Evol. **16**(6), 295–300 (2001)
14. Novozhilov, A.S., Berezovskaya, F.S., Koonin, E.V., Karev, G.P.: Mathematical modeling of tumor therapy with oncolytic viruses: regimes with complete tumor elimination within the framework of deterministic models. Biol. Direct **1**, 6 (2006)
15. Anderson, R.M., May, R.M.: Regulation and stability of host-parasite population interactions: I. Regulatory processes. J. Animal Ecol. **47**, 249–267 (1978)
16. Diekmann, O., Kretzschmar, M.: Patterns in the effects of infectious diseases on population growth. J. Math. Biol. **29**(6), 539–570 (1991)
17. Heesterbeek, J.A., Metz, J.A.: The saturating contact rate in marriage- and epidemic models. J. Math. Biol. **31**(5), 529–539 (1993)
18. Tipping, A.J., Mahon, F.X., Lagarde, V., Goldman, J.M., Melo, J.V.: Restoration of sensitivity to STI571 in STI571-resistant chronic myeloid leukemia cells. Blood **98**(13), 3864–3867 (2001)
19. Bonsall, M.B., Hassell, M.P.: The effects of metapopulation structure on indirect interactions in host-parasitoid assemblages. Proc. R. Soc. Lond. B Biol. Sci. **267**(1458), 2207–2212 (2000)
20. Greenman, J.V., Hudson, P.J.: Host exclusion and coexistence in apparent and direct competition: an application of bifurcation theory. Theor. Popul. Biol. **56**(1), 48–64 (1999)

21. Greenman, J.V., Hudson, P.J.: Parasite-mediated and direct competition in a two-host shared macroparasite system. Theor. Popul. Biol. **57**(1), 13–34 (2000)

22. Hassell, M.P., Bonsall, M.B.: Apparent competition structures ecological assemblages. Nature **388**, 371–373 (1997)

23. Tlsty, T.D., Margolin, B.H., Lum, K.: Differences in the rates of gene amplification in non-tumorigenic and tumorigenic cell lines as measured by Luria-Delbruck fluctuation analysis. Proc. Natl. Acad. Sci. U S A **86**(23), 9441–9445 (1989)

24. Barnes, D.J., Melo, J.V.: Primitive, quiescent and difficult to kill: the role of non-proliferating stem cells in chronic myeloid leukemia. Cell Cycle **5**(24), 2862–2866 (2006)

25. Holyoake, T., Jiang, X., Eaves, C., Eaves, A.: Isolation of a highly quiescent subpopulation of primitive leukemic cells in chronic myeloid leukemia. Blood **94**(6), 2056–2064 (1999)

26. Holyoake, T.L., Jiang, X., Jorgensen, H.G., Graham, S., Alcorn, M.J., Laird, C., Eaves, A.C., Eaves, C.J.: Primitive quiescent leukemic cells from patients with chronic myeloid leukemia spontaneously initiate factor-independent growth in vitro in association with up-regulation of expression of interleukin-3. Blood **97**(3), 720–728 (2001)

Index

N. L. Komarova and D. Wodarz, *Targeted Cancer Treatment in Silico*,
Modeling and Simulation in Science, Engineering and Technology,
DOI: 10.1007/978-1-4614-8301-4, © Springer Science+Business Media New York 2014